CRITICAL THINKING & LOGICAL REASONING WORKBOOK-8

GIFT OF LOGIC™ SERIES

An Essential Resource for Everyone

Boost Your Thinking Skills

Verbal Reasoning
Analytical Reasoning
Pictorial Reasoning

THIRD EDITION

| FOR GRADES 6-12 | STUDENTS, TEACHERS, AND PARENTS |

Ranga Raghuram   **GIFT OF LOGIC™**

Gift Of Logic, Inc

http://www.giftoflogic.com
sales@giftoflogic.com

Critical Thinking and Logical Reasoning Workbook-8
ISBN-13: 978-1494832681
ISBN-10: 1494832682

Third Edition
1-2014

Copyright © 2009 Gift Of Logic, Inc. All rights reserved. No part of this publication may be reproduced, stored in a retrieval system, transmitted in any form or by any means, electronic, mechanical, photocopying, recording or otherwise, without the written permission of the publisher.

License: This book is licensed for use by one person only. Use of this book in a group setting (classroom, workshop, etc) without the written permission of the publisher is prohibited. Unauthorized duplication is strictly prohibited by law. Contact the publisher at sales@giftoflogic.com for classroom/school/group licensing.

GIFT OF LOGIC™
CRITICAL THINKING & LOGICAL REASONING CURRICULUM
12 WORKBOOKS TO BOOST YOUR THINKING SKILLS

For Kindergarten, Grade 1, and Grade 2

Workbook# 0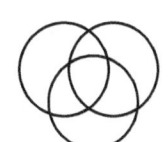

Verbal Reasoning	Finding the truth, Inferencing, Analogies, Synonyms and Antonyms, Agree/Disagree
Analytic Reasoning	Memory drill, Decision making, Positioning, Sudoku
Pictorial Reasoning	Connect the dots, Mazes, Picture Sequence, Spot the difference, etc

Workbook# 1

Verbal Reasoning	Finding the truth, Inferencing, Analogies, Synonyms and Antonyms, Agree/Disagree
Analytic Reasoning	Sorting, Positioning, Picking, Assorted problems, Numeric and Alphabetic Sudoku
Pictorial Reasoning	Picture Sequence, Spot the difference, Odd picture

Workbook# 2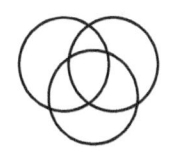

Verbal Reasoning	Finding the truth, Classification, Direct and Inverse relationship, Inferencing, Analogies, Agree/Disagree
Analytic Reasoning	Sequencing, Scheduling, Strategy, Picking, etc
Pictorial Reasoning	Picture Analogy, Odd picture, Pattern matching, etc

For Grade 3, Grade 4, and Grade 5

Workbook# 3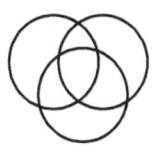

Verbal Reasoning	Not, And, Or, If .. then, Conditional inferencing, Unconditional inferencing, Symbolic Logic
Analytic Reasoning	Lists, Sequencing, Grouping, Venn Diagrams, Graph logic, Number logic, Letter logic, Sudoku
Pictorial Reasoning	Picture sequence, Picture analogy, Odd picture, Picture difference, Pattern matching

Workbook# 4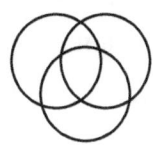

Verbal Reasoning	Contradiction, Converse, Inverse, Contrapositive, Conditional inferencing, Symbolic Logic
Analytic Reasoning	Scheduling, Looping, FIFO, LIFO, Correlation, Venn Diagram, Graph logic, Number logic, Sudoku, etc
Pictorial Reasoning	Picture sequence, Picture analogy, Odd picture, Picture difference, Pattern matching

Workbook# 5

Verbal Reasoning	Biconditional, Categorical inferencing, Cause and Effect, Symbolic Logic, Agree/Disagree, Word and Sentence analogy
Analytic Reasoning	Correlation, Grouping, Venn Diagrams, Graph logic, Number logic, Letter logic, Sudoku, etc
Pictorial Reasoning	Picture sequence, Picture analogy, Odd picture, Picture difference, Pattern matching

********* Essential resource for everyone *********
*http://www.giftoflogic.com *sales@giftoflogic.com

GIFT OF LOGIC™
CRITICAL THINKING & LOGICAL REASONING CURRICULUM
12 WORKBOOKS TO BOOST YOUR THINKING SKILLS

For Grades 6-12, College/University Students, Adults

Primer / Prereq

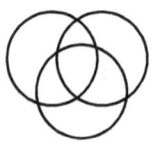

Verbal Reasoning	Logical Operators, Conditional, Categorical and Causal reasoning, Validity, Fallacies, Symbolic Logic
Analytic Reasoning	Positioning, Grouping, Sudoku
Pictorial Reasoning	Pattern perception, Figure formation, Paper folding and cutting, Figure matrix, Rule detection

Workbook# 6

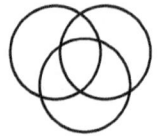

Verbal Reasoning	Arguments-Main point, Must be true, Cannot be true
Analytic Reasoning	Positioning, Grouping, Sudoku
Pictorial Reasoning	Pattern perception, Figure formation, Paper folding and cutting, Figure matrix, Rule detection

Workbook# 7

Verbal Reasoning	Arguments-Strengthening, Weakening
Analytic Reasoning	Positioning, Grouping, Sudoku
Pictorial Reasoning	Pattern perception, Figure formation, Paper folding and cutting, Figure matrix, Rule detection

Workbook# 8

Verbal Reasoning	Arguments - Controversy, Paradox
Analytic Reasoning	Positioning, Grouping, Sudoku
Pictorial Reasoning	Pattern perception, Figure formation, Paper folding and cutting, Figure matrix, Rule detection

Workbook# 9

Verbal Reasoning	Arguments- Assumptions, Reasoning strategy
Analytic Reasoning	Positioning, Grouping, Sudoku
Pictorial Reasoning	Pattern perception, Figure formation, Paper folding and cutting, Figure matrix, Rule detection

Workbook# 10

Verbal Reasoning	Arguments-Flawed reasoning, Analogous reasoning
Analytic Reasoning	Positioning, Grouping, Sudoku
Pictorial Reasoning	Pattern perception, Figure formation, Paper folding and cutting, Figure matrix, Rule detection

********* Essential resource for everyone *********
Get the GIFT OF LOGIC™ today !
*http://www.giftoflogic.com *sales@giftoflogic.com

Dear Reader:

Your decision to purchase this book is commendable. You now have in your hands, a comprehensive, easy-to-read book in Critical thinking and Logical reasoning that will introduce you to three different areas of thinking and reasoning - Verbal, Analytical and Pictorial. Solving problems in Verbal Reasoning is important to develop a critical mind. Solving problems in Analytic Reasoning is important to develop a flexible and resourceful mind. Solving problems in Pictorial Reasoning is important to develop a visually alert mind.

This book is presented in a workbook format to help you progress quickly. Parents and teachers are urged to complete the exercises ahead of the student and assist them whenever necessary with the help of detailed answers provided at the end of the book. This book can be used as a supplementary resource in the regular class room or it can be used during winter and summer vacations. College/ University students, working professionals and retired individuals will also find the Gift Of Logic(tm) Series very useful in enhancing their problem solving abilities, confidence and general intellect.

Critical thinking and Logical reasoning must be practiced consistently to develop strong cognitive skills. After completing the exercises in this book, continue to read the other books in this series to get familiar with different types of Logical reasoning problems.

This workbook is one in a series of twelve workbooks. Please refer to the brochure before this page for a brief description of each workbook. Visit the website http://www.giftoflogic.com for more information.

                                                Happy thinking and reasoning!

TABLE OF CONTENTS

Verbal Reasoning

Controversy..8
Resolving the paradox..31

Analytical Reasoning

Sudoku..50
Positioning...55
Grouping..65

Pictorial Reasoning

Patter perception..75
Figure formation...77
Paper folding and cutting..79

Figure matrix..81
Rule detecn...82

Answers

Verbal..85
Analytic...130
Pictorial...154

Certificate of Completion

Name _____   Date _____

VERBAL REASONING

Name _____   Date _____

CONTROVERSY

In this section on "Controversy", you will develop the ability to analyze two arguments and find out the point at issue (controversy).

You will be presented two arguments that disagree in their conclusions. They may agree or disagree in their premises. A controversy in an argument arises because of the disagreement in the conclusion. The disagreement is also called as the "point at issue". Your task is to pick the correct answer choice that identifies the point at issue.

After the arguments are presented, you will be posed a question as shown below.

 *Which one the following represents the controversy in the arguments?
 *Which one the following is the point at issue?
 *The controversy between Jack and Jill is concerning which one of the following?
 *Martha and Maggie disagree on which one of the following?

Correct answers are those that clearly identify the disagreement/ point at issue in the two conclusions.

Incorrect answers are those that identify agreeing or disagreeing premises or those that do not correctly identify the disagreement in the two arguments.

Name _____ Date _____

| 1 | CONTROVERSY | Sports |

Dad: Stuart likes skating, piano and swimming. So, Stuart should take lessons in at least these three activities.

Mom: I know that Stuart likes skating, piano and swimming. But, he brings a lot of homework everyday and does not have time for three extracurricular activities. So, he can take lessons in at most two activities.

Which one the following represents the point at issue?

A) whether Stuart should take swimming lessons.
B) how many activities Stuart can participate in.
C) whether Stuart should finish his homework in school itself so he can participate in three activities.

Name _____   Date _____

| 2 | CONTROVERSY | Animals |

Cat: People cuddle me five times every day whereas they cuddle you only three times every day. So, they love me more than they love you.

Dog: It is true that they cuddle me only three times every day, but they cuddle me for a longer duration. Therefore, just because they cuddle you more number of times every day does not mean that they love you more.

The controversy between the cat and the dog is regarding which one of the following?

A) who is cuddled more number of times every day.
B) who is loved more.
C) who is cuddled for longer duration.

| 3 | CONTROVERSY | School |

Josh: I like riding my bike to school. It makes me feel good. So, everyone must ride their bike to school.

Dave: I like walking to school. It makes me feel good. So, everyone must walk to school.

Which one of the following is the point at issue between Josh and Dave?
A) whether biking to school is better than walking to school.
B) whether everyone should bike to school or walk to school.

| 4 | CONTROVERSY | Bakery |

Maggie: This cake will taste very good. It was baked with a lot of sugar in it. It also has a layer of vanilla cream on it.

Jackie: My mom too baked a cake last week with a lot of sugar in it. It also had a layer of vanilla cream on it. Baking a cake with lot of sugar does not guarantee that it will taste very good.

What is the point at issue in the debate described above?

A) whether a cake must be baked with lot of sugar or not.
B) whether a layer of vanilla cream is needed or not.
C) whether a cake baked with lot of sugar and topped with vanilla cream will taste good or not.

5 CONTROVERSY — Car

Customer: The brakes in my car are not working properly. I have to press very hard on the brakes and it makes a strange sound when I do that. I bought the car from you. Therefore, you must fix it for free.

Car Dealer: Your car seems to have a unique problem. I have not serviced any car with the type of problem that you describe. Still, I will be glad to fix yours, but your brakes are out of warranty. So, I have to charge you to fix this problem.

What is the point at issue here in the argument between the customer and the car dealer?

A) whether the brakes are working properly or not.
B) whether the problem of the customer's car is unique or not.
C) whether the car dealer should fix the brakes for free or not.

| 6 | CONTROVERSY | Tardiness |

Manager: You have been arriving late to work every day. As a consequence of your tardiness, the progress and morale of your team is being affected negatively.

Employee: My team's morale may be low because of my tardiness, but they have been consistently completing their tasks much before the deadline.

Which one of the following is an issue about which the manager and employee are arguing?

A) whether the employee's tardiness is affecting the team's morale.
B) whether the employee's tardiness is affecting the team's progress.
C) whether the employee is tardy or not.

| 7 | CONTROVERSY | Sales |

Rex: Our competitor, Trojan Computers, has reduced the price of their computers by ten percent. Therefore, they will sell more computers than us.

Tina: Trojan Computers will sell more computers than us only in Asia. The price reduction is effective only in Asia.

Rex and Tina are engaged in a controversy about which one of the following?

A) whether a Trojan computer is better than their own.
B) whether Trojan Computers will sell more computers in all regions.
C) whether the price reduction is in Asia only or not.

| 8 | CONTROVERSY | War and Peace |

Pro-War Activist: The effect of long term conflict is long term suffering of human beings. That is why wars are necessary to put a permanent end to long term conflicts.

Peace Activist: Wars are not effective in putting a permanent end to long term conflicts. After some time, conflicts tend to reappear thus causing continued human suffering. So, we need to find another way to terminate long term conflicts.

The point at issue in the argument between the peace activist and the pro-war activist is

A) whether wars will create new conflicts or not.
B) whether wars can end long term conflicts permanently or not.
C) whether long term human suffering must be ended or not.

Name _____ Date _____

9   CONTROVERSY   Homework

Mom: Diana, after reviewing your homeworks, I have found that you have not completed homeworks A,P, R,M and Z.

Diana: Mom, please review it again, I have completed homeworks A, R and Z, but not the others.

Mom and Diana disagree about which one of the following statements?

A) Diana has not completed assignments A,R and Z.
B) Diana has not completed assignments P and M.

Name _____ Date _____

10   CONTROVERSY   Law

Judge: You have committed an offense. You have stolen money from the shopkeeper. You also stole his watches and utensils. So, you deserve the maximum punishment.

Defendant: I stole only a few dollars from the shopkeeper. I did not steal his watches and utensils. Since this is only a minor offence, I deserve the minimum punishment only.

Which one of the following expresses the point at issue between the defendant and the judge?

A) the amount of punishment given for minor offense.
B) the amount of punishment that must be given to the defendant.
C) whether committing an offense is immoral.

| 11 | CONTROVERSY | Service |

Principal: Kazaki is the best sportsman this year. His academic record is also good with a score of ninety percent in all the subjects. He also did excellent service to assist the disabled people. Therefore, considering all the merits, without doubt, no one but Kazaki must be awarded the gold medal this year.

Assistant Principal: Kazuaki also deserves the gold medal this year. He was an important player in the school sports team. His academic score of ninety three percent is better than Kazaki's. Kazuaki also has done social service to help the poor and the disabled.

The controversy in the argument between the principal and the assistant principal is

A) whether social service must be considered while deciding who should get the gold medal.
B) whether the gold medal must be shared by both Kazaki and Kazuaki.

12   CONTROVERSY   Painting

Rachel: All the artwork submitted to this year's art festival are of very high standard. The oil paintings are beautiful. The acrylic paintings are stunning. The watercolor paintings are mind boggling. So, these artworks are great examples for a beginner to learn from.

Shannon: The acrylic paintings are certainly breathtaking. However, the oil paintings and the watercolor paintings seem to be of good standard, but not of very high standard as you describe. I am not sure that these artworks are worthy of mimicry by beginners.

Rachel and Shannon disagree about which one of the following?

A) that the acrylic paintings are very impressive.
B) that oil paintings are of good standard.
C) that beginners can learn from the artwork submitted to the festival.

13 CONTROVERSY — Cartoons

Mom: Donna, as you know, you have been watching cartoons for more than four hours today. This will hurt your mental health and make you dream about strange things. Therefore, you must stop watching cartoons now.

Donna: I read today in the Journal of Health that watching up to six hours of cartoons per day is the limit, that when exceeded, will cause delirium. So, I can watch cartoons for a bit longer.

Mom and Donna disagree on which one of the following?

A) whether watching cartoons beyond a limit will cause mental problems.
B) how long one can watch cartoons before experiencing mental problems.

14   CONTROVERSY   Alternate

King A: My soldiers have captured yours. Hence, you must surrender to end the war now.

King B: My soldiers are still fighting yours. The war is still going on. Therefore, my solders are not in your command.

King A: Your chief of army has run away from the battlefield. That is why your soldiers have surrendered to my command.

King B: My chief of army has run away, but my soldiers will fight to the end.

King B disagrees with King A on which of the following?

A) that his chief of army ran away from the battlefield.
B) that the war is already over.
C) that his soldiers have surrendered.

| 15 | CONTROVERSY | Books |

Elizabeth: I like electronic versions of books over printed versions.

Paul: I don't like electronic versions. You need a computer to read them.

Elizabeth: That is exactly why I like them. I can read them wherever I go and don't need to remember to take the books with me because it is always in my computer.

Paul: Well, I like printed books because, I can highlight important words whereas you cannot do that on electronic versions. Moreover, I don't like to stare at a computer screen while reading books. But, I like the fact that you do not have to remember to take electronic books with you.

Elizabeth: Electronic books also let you highlight words and write your own notes.

Based on the conversation above, Elizabeth and Paul disagree on which one of the following?

A) that one does not have to remember to take electronic books with them.
B) that highlighting can be done in electronic versions of books.

| 16 | CONTROVERSY | Energy |

Mayor: Oil is a natural resource that is getting depleted quickly. We must encourage the use of alternate sources of energy such as wind and solar energy. Using up our oil resources quickly will be catastrophic and so it is a good idea to penalize industries that depend only on oil.

Governor: Exploring alternative sources of energy is critical. We should consider giving tax rebates to industries for using solar and wind energy. This plan will help us conserve oil for future use. But, since penalizing industries that depend on oil only could have a negative effect on our economy, we must not do so.

The mayor and the governor disagree on which of the following?

A) that oil is a scarce resource.
B) that wind and solar energy are not alternative sources of energy.
C) that penalizing industries for using oil is a good idea.

| 17 | CONTROVERSY | Internet |

Amanda: Purchasing books from the Internet is very convenient and safe. You don't need to go to a bookstore. You can pay with a credit card anytime. The books get shipped to you within days of placing the order.

Vivek: Ordering books from the Internet from the comfort of your home is very convenient, but it is not always safe. The books that you order will arrive at your house quickly. However, your credit card information can be stolen if the Internet site from where you are purchasing does not have proper security features.

Vivek and Amanda are in disagreement over which of the following?

A) that books take a lot of time to get delivered when ordered from the Internet.
B) that you have to go to the book store to place the orders from Internet.
C) that purchasing books from the Internet is always safe.

18   CONTROVERSY   Sports

Coach: If you observe the performance of soccer players in our school team, all those that were coached since they were five years old play significantly better than those that were not. This goes to prove that coaching for soccer should begin at five years of age.

Principal: If you observe the performance of soccer players in all the schools in our area, you will note that all those who started taking soccer lessons from ten years of age play as well as those who started taking lessons when they were five years old. So, they don't have to necessarily start learning soccer from five years of age.

The Principal and Coach disagree on which of the following?

A) whether only those players who are coached play well.
B) whether soccer players must be coached from five years of age.

19   CONTROVERSY   Bus

Bus Driver: Sir, your are sitting in a seat that is reserved for the handicapped. You need to vacate it immediately.

Passenger: Since, there is no handicapped person in the bus now, I am justified in being seated here until one shows up.

Bus Driver: The rule is that, even if there is no handicapped person in the bus, the seat must not be occupied by the non-handicapped.

The bus driver and the passenger disagree on which of the following?

A) that a seat reserved for the handicapped is meant to be occupied by handicapped people.
B) that a seat reserved for the handicapped must only be occupied by the handicapped at all times.
C) that handicapped people can sit in seats not reserved for them.

20 — CONTROVERSY — Customer

Customer: The vacuum cleaner that I purchased from your store is defective. It does not suck the dust quickly. Moreover, it makes a strange noise that I do not like.

Technician: These vacuum cleaners make a distinct noise that you may not like, but they are normal. Moreover, the motor inside is not very powerful, and so they do not suck the dust quickly, but this too is normal.

The customer and the vacuum cleaner technician disagree on which of the following?

A) that the vacuum cleaner makes noise when turned on.
B) that the vacuum cleaner is defective.
C) that it does not suck the dust quickly.

21   CONTROVERSY   Medical

Doctor Stone: The patient must be given a heavy dose of sedatives. He is very agitated about something.

Doctor Stewart: The latest bulletin from the head of our department advises that patients must not be given a heavy dose of sedatives when the only symptoms displayed are related to agitation. This patient had a prolonged argument with his wife. I don't see any other symptom besides an agitated mind. It is preferable to give him a light dose of sedatives to calm his nerves.

Doctor Stone and doctor Stewart are in disagreement over which of the following issues?
A) whether the patient must be given a heavy dose of sedatives.
B) whether the patient is agitated or not.

22   CONTROVERSY   Pharmacy

Pharmacist A: The Doctor has prescribed two tablets for this patient to be taken simultaneously. There is a new syrup available that has the same strength as these two tablets combined together. So, we can give this new syrup to this patient instead of the two tablets.

Pharmacist B: As a matter of policy, the medicines must be dispensed exactly as prescribed regardless of their form. There should be a specific reason why he has not prescribed the syrup.

The pharmacists agree on the following except

A) that the doctor has prescribed two tablets when a syrup of equal strength is available.
B) that medicines can be dispensed in any form as long as their strengths are the same.

23 CONTROVERSY — Landlord

Tenant: Painting the inside of the house is not a violation of the lease agreement as you allege, since this is clearly allowed as per the clauses in the lease agreement.

Landlord: The lease agreement states that whenever you paint the inside of the house, you must also paint the outside of the house, which you obviously failed to do. That is the reason why you are in violation of the lease agreement.

The Tenant and the Landlord disagree on which of the following?

A) whether there has been a violation of the lease agreement.
B) whether the outside of the house must be painted whenever the inside of the house is painted.

Name _____   Date _____

RESOLVE THE PARADOX

In this section on "Resolve the Paradox", you will develop the ability to identify and resolve a paradox. A paradox is also referred to as a contradiction or a puzzle.

A paradox is a set of facts that are in contradiction with each other. Paradoxes begin with premises that are intended to lead us to a logical conclusion. This is the first part of the paradox. Suddenly, the paradox will shift its line of reasoning and state a contradictory set of facts, thus creating a puzzle. This is the second part of the paradox. A paradox uses counter premises such as "however", "nevertheless", "but", "although", etc to describe this shift in direction. Your task is to find a statement that would explain/resolve the contradiction in the paradox. The structure of a paradox can be thought of as follows:
  (premises in first part) ↔ (counter premise) ↔ (premises in second part)

After the paradox is presented, questions similar to the ones below are posed:
 * Which one of the following statements would most effectively resolve the paradox?
 * Which one of the following would help to reconcile the conflict in the passage above?

A correct answer is one that explains/resolves the contradiction.

Incorrect answers are those that do not explain the contradiction at all or those that explain the paradox only partially.

| 1 | RESOLVE THE PARADOX | Sports |

Ronaldo can score three goals in any soccer game that he plays. He is an exceptional player and practices everyday to maintain his performance. But, in the crucial final game yesterday against the Panthers, he scored only two goals.

Which one of the following, if true, explains this paradox?

A) In the final game, his team played against the Panthers, a much stronger team.
B) Ronaldo does not like to play against the Panthers since they wear a red color jersey.
C) Ronaldo was not feeling well during the final game.

Name _____   Date _____

2   RESOLVE THE PARADOX   Traffic

Trucks carrying hazardous materials are not allowed to enter the city of Greenville. They must take a detour around the city. The traffic in the city is very congested and it cannot afford to risk any accident involving hazardous trucks. But, the traffic control room received a call a few minutes ago that several trucks carrying benzene, a hazardous chemical, were inside the city limits.

Which one of the following, if true, most helps to explain the paradox in the passage?

A) motorbikes too are not allowed into the city of Greenville, but you can see them everywhere.
B) hazardous trucks often enter the city in order to save thirty minutes of time driving on the detour route.
C) the detour route has been closed due to an accident that occurred a few minutes ago and the traffic police have authorized the hazardous trucks to enter the city limits temporarily.

| 3 | RESOLVE THE PARADOX | Crime |

Fingerprints are the patterns found at the edge of our fingers. These are unique to each human being and therefore are used as absolute proof of identity. So, the police were very puzzled that even though the video camera at the store where a theft occurred shows a clear picture of Travis stealing with his own hands, his fingerprints were not found anywhere in the store.

Which one of the following, if true, explains the puzzle about Travis's fingerprint?

A) Travis is a very religious person and hence would definitely not have stolen from the store.
B) Travis wore gloves while stealing to avoid leaving any fingerprints.
C) Travis had not touched anything when the alarm went off and he left without leaving any fingerprints.

Name _____   Date _____

| 4 | RESOLVE THE PARADOX | Photography |

Sophia bought a very expensive camera because she was not satisfied with the quality of pictures produced by cheap cameras. The expensive camera has a very advanced technology that produces crystal clear pictures. She used her new camera during her visit to New York, but when she developed the pictures taken with it, she was disappointed that they were of the same poor quality as the ones produced by the cheap camera that she once owned.

Which one of the following best explains the paradox described above?

A) Sophia is not a professional photographer and so it does not matter whether she uses a cheap camera or an expensive camera.
B) Sophia has the misguided perception that an expensive camera will yield better pictures than an inexpensive one.
C) The new camera was not configured to take advantage of its advanced technology.

| 5 | RESOLVE THE PARADOX | Tornado |

Hundreds of volunteers and police were sent to the city of Oceana to help people recover from the recent tornado. They helped the people to rebuild their homes and roads and bridges. Nevertheless, there is widespread unemployment in the city.

Which one of the following, if true, will explain the widespread unemployment in the city of Oceana?

A) The volunteers did not help rebuild the businesses in the city.
B) The nearby city of Sienna was also hit by the tornado and they too have widespread unemployment.
C) Unemployment is not only in the city of Oceana, it is prevalent everywhere in the country.

6 RESOLVE THE PARADOX — Farmer

Farmer Joe purchased tomato seeds from the Greenhouse Nursery which carries seeds of the best quality. The plants that grow from these seeds yield tomatoes that are large, red and delicious. Farmer Joe planted these seeds in several acres at his farmland and eagerly waited to taste the large tomatoes. When the tomatoes grew, even though they were red and delicious, farmer Joe was disappointed that they were small.

Which one of the following, if true, most helps to resolve the mystery behind the incorrect size of the tomatoes?

A) Farmer Joe should have purchased the seeds from Evergreen Nursery which carries better quality seeds than Greenhouse Nursery.
B) Other farmers bought the exact same seeds and obtained large, red and delicious tomatoes.
C) Farmer Joe did not apply the exact amount of fertilizer that was recommended by Greenhouse Nursery for growing large tomatoes.

7 RESOLVE THE PARADOX — Currency

A fitness club has two types of memberships – the adult only "Single" membership and the "Family" membership for families with adults and children. The neighborhood where this club is located has more families than single people. The owner of the club is puzzled as to why an overwhelming majority of his customers do not have the "Family" membership.

Which one of the following, if true, would help to resolve the puzzle experienced by the owner of the fitness club?

A) The fitness club conducts several free fitness programs specially for children.
B) The five feet deep swimming pool is unsuitable for most children.
C) Most people who are eligible for the "Single" membership do not like to go to a fitness club that has family memberships.

| 8 | RESOLVE THE PARADOX | Peak hour |

Traffic in the Interstate Highway #75 has been a nightmare for peak hour commuters. To address this problem, the Department of Transportation opened an additional lane in the highway. They estimated that it would take only thirty minutes to travel ten miles in the freeway now, whereas it used to take forty five minutes before. But, much to their surprise, the matter has become worse and it now takes almost an hour to travel ten miles in the freeway.

Which one of the following, if true, will help resolve the paradox described above?

A) People driving in the freeway after peak hours take only twenty minutes to travel ten miles in the freeway.
B) New commuters have started using the freeway to take advantage of the additional lane.
C) The speed limit during peak hours was increased to relieve congestion.

9. RESOLVE THE PARADOX — Health

Rhonda saw a doctor to treat a skin infection that causes itching in her hands. The doctor prescribed her an ointment that has provided relief within two weeks to many that had a similar problem. Rhonda followed the doctor's instruction and applied the ointment twice a day, but much to her disappointment, the itching caused by the infection did not go away even after two weeks.

Which one of the following, if true, will help to explain Rhonda's disappointment?

A) Rhonda used the ointment on her legs last year and the itching caused by the skin infection went away within two days.
B) The ointment created an adverse reaction when it interacted with the metallic ring in her fingers.
C) Rhonda washed her hands three times a day to help prevent skin infections.

10 RESOLVE THE PARADOX — Tutor

Molly's mom hired a private tutor to help with her school lessons. Molly has been slipping behind in her academic performance and her parents thought that having some help from a tutor would help improve her performance. But, much to their dismay, Molly's performance continued to decline.

Which one of the following, if true, would help resolve the paradox?

A) Molly's tutor taught her more than what was necessary.
B) Molly did not take the tutorials seriously.
C) Students with private tutors show a significant improvement in their academic performance.

11 RESOLVE THE PARADOX — Real Estate

The Creek Hollow community of houses was built five years ago. The community has the benefit of being situated close to shopping malls and good schools. Because of this benefit, in the last five years, houses in the community sold out quickly when they came up for sale. Surprisingly, it now takes twice as much time to sell a home in this community as it took five years ago.

Which one of the following, if true, most helps to explain the paradox described above?

A) The home owners have decorated the entrances to Creek Hollow with pretty flowers.
B) Lot of young couples have purchased homes in the community.
C) Due to lack of jobs in the city where the Creek Hollow community is located, the number of home buyers have significantly reduced.

12 RESOLVE THE PARADOX — Entertainment

The race for the "Miss Galaxy" award has boiled down to three contestants – Nicole, Anita, and Salma. Each of them qualified by virtue of being "Miss Country" of their respective countries and have won several rounds to come to the final competition. The award function was televised all over the world. The viewers of the program chose Nicole for the award through internet-voting, but much to their disbelief, Anita was crowned as the "Miss Galaxy" for this year.

Which one of the following, if true, best explains the reason for this puzzling decision regarding the recipient of the "Miss Galaxy" award?

A) Internet is accessible to a majority of people in Anita's country.
B) People in Nicole's country have the most access to internet.
C) The viewer's internet-vote was the only factor that decided the winner.
D) The viewer's internet-vote played only a minor role in the selection of the winner.

13   RESOLVE THE PARADOX   Entertainment

A 24-theater mega movie complex was opened recently in a city where surveys have shown that people like to see one movie every month. The owner of the complex, Mr. Strongman, was hopeful of making a good profit from ticket sales. But, much to his disillusionment, the number of visitors to his complex was much lower than his estimates and he is running the complex at a loss.

Which one of the following, if true, most helps to explain the low ticket sales in the movie theater complex?

A) There is a trash-collecting facility near the theater complex.
B) The employees of the theater complex do not like Mr. Strongman.
C) There are already several well established movie theater complexes in the city.

14 — RESOLVE THE PARADOX — Candy

Dad: There were ten candies in the box and I asked you to leave two for your brother. Now, I see that there is only one candy in the box. What happened to the other candy?

Alex: I ate only eight candies. I wrote on my hand the number of each of the eight candies as soon as I ate them – you can see the numbers from 1 to 8 on my hand. I don't know why one candy is missing.

Which one of the following, if true, most helps to reconcile Alex's response to his Dad's question?

A) After eating eight candies, Alex put the remaining two in the box.
B) There were no holes in the box that held the candies.
C) Alex did not notice that the eighth and ninth candies were attached to each other.

15   RESOLVE THE PARADOX   Firefighter

Home-Owner: A very intense fire started in the bedroom as soon as I turned on the room heater. I knew that I needed help to deal with it and that is why I called the fire station.

Firefighter: Yes, I can see that the fire was very intense indeed, but you should have called us a lot sooner. Nevertheless, our inspection of your house indicates that there is very little damage to your house.

Which one of the following, if true, most helps to explain why there was very little damage to the house?

A) The intensity of the fire caused expensive damage to the bedroom and the adjacent living room.
B) Metal fragments from the burned out room heater were found to be at a temperature of 500 degrees Fahrenheit.
C) The intensity of the fire turned on an automatic sprinkler system that doused the flames.

| 16 | RESOLVE THE PARADOX | Hurricane |

Several families with school children were displaced due to very powerful hurricanes that lashed the coastal areas. The Kindness Elementary School admitted two hundred of these displaced students. Since this additional load was very significant, there was concern about the quality of education that each student would receive. But, much to everyone's surprise, the annual test results proved that the initial concern about the quality of education was unfounded.

Which one of the following, if true, most helps to explain why the quality of education did not deteriorate because of the additional student load?

A) Seven new teachers were hired to help the current teachers.
B) The student teacher ratio was maintained at the same level after the hurricane as it was before the hurricane.

17 RESOLVE THE PARADOX — Bulbs

Four new 60 Watt bulbs, all made by the same company, were turned on at the exact same time to test their longevity. It was expected that they will all burn out at the exact same time. But, much to everyone's surprise, they all burned out several days apart.

Which one of the following, if true, will resolve the paradox presented above?

A) Just like human beings born on the same day don't die on the same day, bulbs too have different life spans.
B) It is not realistic to expect four new, exactly same type of bulbs to have the same longevity.
C) The vacuum-creating process, a key factor deciding the longevity of bulbs, is not a perfect process.

Name _____ Date _____

ANALYTICAL REASONING

Name _____ Date _____

1   SUDOKU

Solve the following Sudoku. A correctly solved Sudoku has numbers 1-9 appearing only once in each row, each column and each 3x3 grid. Solving Sudokus will help you to gain valuable analytic skills.

3	5	4		9	1	8	2	7
			4	5	7	9	3	1
1	9	7	3	8			5	6
6	1	3	2	7	4	5		8
	8	9				2	6	4
2	4	5	9	6	8	7	1	
9	2		7	4	6	3	8	5
	3	8	1	2	5			
	7	6	8	3	9	1	4	2

Analytical Reasoning   Answers-125
© Gift Of Logic, Inc * Copying prohibited

2
SUDOKU

Solve the following Sudoku. A correctly solved Sudoku has numbers 1-9 appearing only once in each row, each column and each 3x3 grid. Solving Sudokus will help you to gain valuable analytic skills.

			7	9	5	6	4	2
4	7	9		3	2	1	5	8
2	6	5	1	4	8	3	7	9
3	5	1					2	6
7		6	2		4	5		3
8	4	2	5	6	3	7		1
6	8	3	4	2	7			
5		7	9		6	2		4
9	2	4	3	5	1	8	6	7

Analytical Reasoning   Answers-126
© Gift Of Logic, Inc * Copying prohibited

3

SUDOKU

Solve the following Sudoku. A correctly solved Sudoku has numbers 1-9 appearing only once in each row, each column and each 3x3 grid. Solving Sudokus will help you to gain valuable analytic skills.

	3	5	1	2	9	7	4	
9		1	4	6	5	2		8
2	6		3	7	8		5	9
3	1	9		8		5	7	4
4	8	2	5		7	6	9	3
7	5	6		3		8	2	1
5	4		6	9	2		1	7
6		3	7	4	1	9		5
	9	7	8	5	3	4	6	

Analytical Reasoning

4 SUDOKU

Solve the following Sudoku. A correctly solved Sudoku has numbers 1-9 appearing only once in each row, each column and each 3x3 grid. Solving Sudokus will help you to gain valuable analytic skills.

	7	3	8	6	1		2	4
1		4	3		5	6	9	7
5	2		7	4	9	1	8	3
7	9	5		3			4	1
3	1	8	2		4	7		5
4		2	5	1	7	8		9
6	3	1	4		2	9		8
2		9	1	8	3		7	6
8	4	7			6	3	1	2

Analytical Reasoning

Name _____ Date _____

| 5 | | | | SUDOKU | | | | |

Solve the following Sudoku. A correctly solved Sudoku has numbers 1-9 appearing only once in each row, each column and each 3x3 grid. Solving Sudokus will help you to gain valuable analytic skills.

1		2	8	4	5	6	3	
3	4			2	6	5		1
6	5	8	9		1		4	7
2	1	4	6			3	7	5
7	8	6	2			1	9	4
5	9	3		1	7		2	6
4	3		5	9	8	7		2
9		5	3	7	2	4		
	2	7	1	6	4	9	5	3

Name _____   Date _____

POSITIONING

1

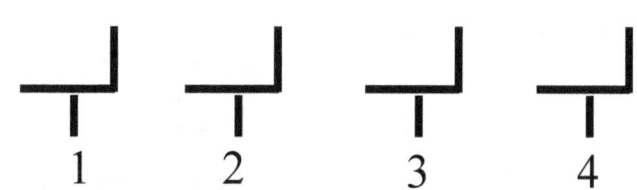

Two boys and two girls must be seated in the chairs shown above.

1) How many seating configurations are possible? Write them below.

_____

2

Two boys and two girls must be seated in the chairs shown above. A boy and a girl must sit next to each other.

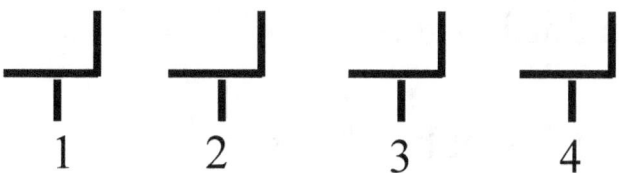

1) If a boy sits in the fourth chair, another boy can sit in the first chair.
   A) True   B) False

POSITIONING

3 next to, right of

| 1 | 2 | 3 | 4 |

Place the following animals in the four spots shown according to the conditions listed below: Zebra, Horse, Elephant, Monkey

Zebra must be in the first spot.
Monkey must be next to the Zebra.
Horse must be to the right of the Elephant.

---

4 immediately, fixed position

| 1 | 2 | 3 |

Three birds, a Parrot, a Peacock, and an Eagle must be placed in three cages shown according to the rules described below.

Parrot must be immediately to the left of Peacock.
Peacock must be in the third cage.

1) In which cage can the Parrot be placed?
    A) cage# 1   B) cage# 2

2) In which cage can the Eagle be placed?
    A) cage# 1   B) cage# 2

Name _____   Date _____

POSITIONING

5   right of

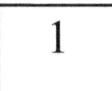

Four balls, yellow, red, blue, and green in color, must be placed in the boxes shown above with only one ball in each box. The yellow ball must be in box# 1. The blue ball must be to the right of the red ball.

1) Can the blue ball be placed in box# 2?
   A) Yes   B) No

2) If the blue ball is placed in box# 3, where can the red ball be placed?
   A) box# 2   B) box# 4

3) If the blue ball is in box# 4, where can the red ball be placed?
   A) box# 2   B) box# 3   c) box# 2 or box# 3

Name _____ Date _____

POSITIONING

6 inclusive or

| 1 | 2 | 3 | 4 |

Four plants are to be planted in four spots shown above.
Jasmine must be planted in spot 1 or spot 3, but nowhere else.
Rose must be planted in spot 2 or spot 4 only.

1) Which of the following plantings is correct?
    A) spot-1: jasmine spot-2: rose spot-3: rose spot-4: jasmine
    B) spot-1: jasmine spot-2: rose spot-3: jasmine spot-4: rose

---

7 exclusive or

| 1 | 2 | 3 |

Three spots shown above are available for planting.
Either carrot or tomato can be planted in spot# 1, but not both.
Either tomato or peach can be planted in spot# 2, but not both.
Either peach or carrot can be planted in spot# 3, but not both.

1) If tomato is planted in two spots, then carrot cannot be planted.
    A) True   B) False

2) If tomato is planted in spot# 2 and carrot is planted in spot# 3, then peach cannot be planted.
    A) True   B) False

POSITIONING

8 or

| 1 | 2 | 3 | 4 |

The four boxes shown above are to be painted with one color each.
Boxes 1 and 3 must be painted red.
Box# 2 can be painted green or blue.
Box# 4 can be painted blue or yellow.

1) Which of the following choice of colors is valid?
    A) red   blue   green   blue
    B) red   green   red   yellow
    C) red   blue   red   green
    D) red   blue   red   blue

---

9 end, and

| 1 | 2 | 3 | 4 |

The four boxes shown above are to be painted with one color each.
The boxes at the ends must be painted red and green.
Box# 2 must be painted blue and box# 3 must be painted yellow.

1) Which of the following choice of colors is valid?
    A) red, blue, yellow, green
    B) green, blue, yellow, red
    C) red, blue, green, yellow
    D) green, red, blue, yellow

POSITIONING

10

vacancy

Gina, Gemma, and Gordon sit in three chairs positioned as shown. The fourth chair is vacant.

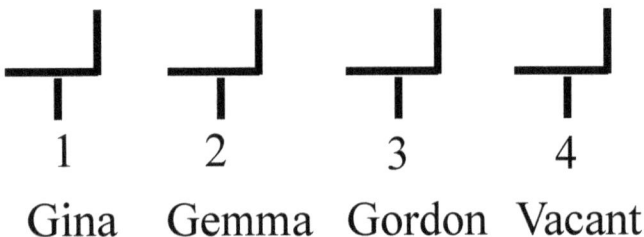

    1   2   3   4

  Gina Gemma Gordon Vacant

1) If Gordon moves to position 4, which chair will become vacant?
   A) 3   B) 4

2) If Gemma and Gordon each move one chair to the right, which chair will become vacant?
   A) 2   B) 3

3) Which of the following moves will make the chair next to Gina vacant?
   A) Move Gordon to chair #4
   B) Move Gemma to chair #4

POSITIONING

11 corners

Nancy, Nina, Nick, and Noor are to be seated in four chairs 1, 2, 3 and 4 respectively. The chairs are placed in one single row as shown below.

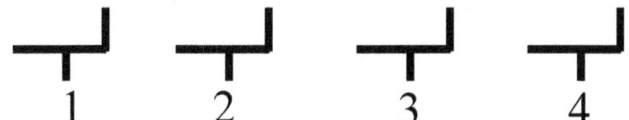

Nina and Nick must sit at the chairs at the corners.
Noor must sit to the left of Nancy.

Based on the scenario described above, answer the following questions.

1) Nancy must sit in the third chair.
    A) True   B) False

2) Nina must be sit in the first chair.
    A) True   B) False

3) Nick must sit in the fourth chair.
    A) True   B) False

4) Nina must always sit to the left of Noor.
    A) True   B) False

5) Noor cannot sit in the third chair.
    A) True   B) False

POSITIONING

12 Harry, Hilga, Hughes, and Heather are to be seated in four consecutive positions, one person only in each position. Harry can be in position 1 or 4 only. Hilga and Heather must sit next to each other.

1) If Hughes sits in position 1, Harry can sit in which of the following positions?
    A) 2   B) 3   C) 4

2) If Harry is in position 1 and Hughes is in position 4, Hilga must be in position 2.
    A) True   B) False

---

13 Ian, Irene, Iqbal, and Ishwar are to be seated in four consecutive chairs numbered as 1, 2, 3, and 4.

Either Ian or Irene can sit in the first chair. Either Irene or Iqbal can sit in the second chair. Either Iqbal or Ishwar can sit in the third chair. Either Ishwar or Ian can sit in the fourth chair.

1) If Ian sits in the first chair, Ishwar can sit in which of the following chairs?
    A) 3   B) 4

2) If Ian sits in the fourth chair, Irene can sit in which of the following chairs?
    A) 1   B) 4

3) If Iqbal sits in the third chair, Ian can sit in the fourth chair.
    A) True   B) False

POSITIONING

14 not next to

A parrot, a butterfly, an eagle, and a flamingo are to be seated in four spots positioned as shown below.

        1   2   3   4

The butterfly must not be next to the eagle. The parrot must not be in the first spot.

1) Which of the following is a correct positioning of the birds?
    A) Parrot, Butterfly, Eagle, Flamingo
    B) Butterfly, Parrot, Eagle, Flamingo

---

15 not next to

A dog, a cat, an elephant, and a lion are to be seated in four spots positioned as shown below.

        1   2   3   4

The Dog must be in first spot.
The Elephant must not be next to the dog.
The Lion must be to the right of the elephant.

1) The Cat must be in which spot?
    A) 2 B) 3 C) 4

2) The Elephant cannot be in spot 4.
    A) True   False

POSITIONING

16 before, after

Nancy and Olivia must sit in a row along with other people according to the following rules.

Nancy must sit before Olivia.
Nancy must sit immediately after Mary.

1) Which of the following seatings is correct based on the above rules?
    A) Pat, Mary, Olivia, Nancy
    B) Mary, Nancy, Pat, Olivia

---

17 not before, not after

Roy and Uday must stand in a row along with other people according to the following rules.

Roy cannot stand before Sam.
Uday cannot stand after Tom.

1) Which of the following positioning is correct?
    A) Sam, Rohan, Uday, Tom
    B) Sam, Rohan, Tom, Uday
    C) Rohan, Uday, Sam, Tom

Name _____   Date _____

| 1 | GROUPING | any two |

SCENARIO

Three boys, Ali, Bob, and Chen meet to play chess. Only two people play chess at any given time.

QUESTIONS

1) How many chess games must be played so that each boy plays the other. Write the names of the players in each game.
   Game# 1 :
   Game# 2 :
   Game# 3 :

| 2 | GROUPING | if-then |

SCENARIO

Three girls Asha, Brandi, and Christina went to watch a movie, but only two seats were available. If Asha watches the movie, then Brandi must also watch the movie.

QUESTIONS

1) Which of the following pairs can watch the movie?
   A) Asha, Brandi
   B) Brandi, Christina
   C) Christina, Asha

Name _____ Date _____

| 3 | GROUPING | any two, if and only if |

SCENARIO

Three animals, an Anteater, a Bison, and a Cheetah decide that two of them must go for a walk. If and only if the Anteater goes for a walk will the Bison also go for a walk.

QUESTIONS

1) Which of the following pairs of animals can go for the walk?
    A)   Anteater, Bison
    B)   Bison, Cheetah
    C)   Cheetah, Anteater

| 4 | GROUPING | any two, if-then |

SCENARIO

Three animals, an Anteater, a Bison, and a Cheetah decide that two of them must go for a walk. If the Anteater goes, then the Bison cannot go.

QUESTIONS

1) Which of the following pairs of animals can go for the walk?
    A)   Anteater, Bison
    B)   Bison, Cheetah
    C)   Cheetah, Anteater

Analytical Reasoning   Answers-147
© Gift Of Logic, Inc * Copying prohibited

| 5 | GROUPING | any three from four |

SCENARIO

Three cakes need to be selected from a set of four cakes, named A, B, C, and D.

QUESTIONS

1. Write the valid combinations of cakes that can be selected.

| 6 | GROUPING | any three from four, condition |

SCENARIO

Three cakes need to be selected from a set of four cakes, named A, B, C, and D.   Cakes B and C must not be selected together.

QUESTIONS

1. Which of the following are valid combinations of cakes that can be selected.
- A)   A,B,C
- B)   C,B,D
- C)   C,D,A
- D)   D,A,B

| | | GROUPING | two groups, condition |

SCENARIO

Four students, Amber, Brian, Calvin, and David, need be split into two teams, the Blue team and the Green team. There should be exactly two students in each team.

If Amber is in the blue team, then Brian must be in the Green team.
If Calvin is in the blue team, David must be in the Green team.

QUESTIONS

1) Which of the following team selections are possible based on the above scenario?

Blue team	Green team	Possible?
Amber, Brian	Calvin, David	
Amber, David	Calvin, Brian	
Amber, Calvin	Brian, David	
Calvin, David	Amber, Brian	

8   GROUPING   same group size

SCENARIO

Four students, Prince, Queenie, Raj, and Sam are to be sent into one of two rooms, the Blue room or the Green room. There should be exactly two students in each room.

Queenie should be in the green room.
Raj should be in the blue room.

QUESTIONS

1) Analyze each row and write Yes/No in the column marked 'Possible', if the assignment of the students to the two rooms is valid.

Blue room	Green room	Possible?
Prince, Queenie	Raj, Sam	
Sam, Raj	Queenie, Prince	
Queenie, Raj	Prince, Sam	
Raj, Prince	Sam, Queenie	

| 9 | GROUPING | two in same group |

SCENARIO

Four students, Amber, Brian, Calvin, and David need be sent into two rooms, either to the Blue room or to the Green room. There should be exactly two students in each room.

Amber and Calvin must be in the same room.
Brian and David must be in the same room.

QUESTIONS

1) Which of the following can represent valid room occupancy based on the above scenario?

Blue room	Green room	Possible?
Amber, Brian	Calvin, David	
Amber, Calvin	Brian, David	
Brian, Calvin	Amber, David	
Brian, David	Amber, Calvin	

Analytical Reasoning

10 GROUPING — not in the same group

SCENARIO

Four professors, A, B, C, and D are to be seated in two tables, two in each table. Professors B and C cannot be seated at the same table.

QUESTIONS

1) Which of the following can represent valid table occupancy based on the above scenario? Write Yes/No in the "Possible" column.

Blue table	Green table	Possible?
A, B	C, D	
B, C	A, D	
D, A	C, B	

Name _____   Date _____

11 GROUPING   or

SCENARIO

Four professors, A, B, C, and D are to be seated in two tables, a green one and a blue one. Each table must seat two professors. Professor A or professor B can sit in the blue table, but not both.

QUESTIONS

1) Which of the following can represent valid table occupancy based on the above scenario?

Blue table	Green table	Possible?
A, B	C, D	
A, D	B, C	
B, C	A, D	

12 GROUPING   only, or

SCENARIO   Four professors, A, B, C, and D are to be seated in two tables, a green one and a blue one. Each table must seat two professors. Only professor A or professor B can sit in the blue table, but not both. Only professor B or professor C can sit in the green table, but not both.

QUESTIONS

1) Which of the following can represent a valid table occupancy based on the above scenario?

Blue table	Green table	Possible?
D, A	C, B	
C, B	B, D	
B, A	C, D	
B, D	C, A	

Analytical Reasoning   Answers-152
© Gift Of Logic, Inc * Copying prohibited

13 GROUPING and

SCENARIO

Four professors, A, B, C, and D need to be seated in two tables marked Blue and Green respectively. Each table must seat two professors. Professor A must sit in the blue table and professor B must sit in the green table.

QUESTIONS

1) Which of the following can represent valid table occupancy based on the above scenario?

Blue table	Green table	Possible?
B, A	C, D	
C, A	B, D	
C, D	A, B	
D	B, C	

14 GROUPING or

SCENARIO

Four professors, A, B, C, and D need to be seated in two tables marked blue and green respectively. Each table must seat two professors. Professor A must sit in the blue table or Professor B must sit in the green table.

1) Which of the following can represent valid table occupancy based on the above scenario?

Blue table	Green table	Possible?
A, B	C, D	
C, D	A, B	
A, C	B, D	
B, D	A, C	

Analytical Reasoning   Answers-153

Name _____ Date _____

PICTORIAL REASONING

Name _____   Date _____

PATTERN PERCEPTION - MISSING PATTERN

Find the correct figure from the three alternatives given that will fit logically into the missing portion of the figure on the left.

1 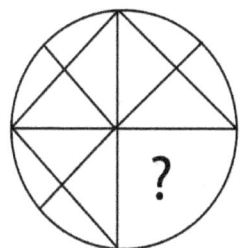 A 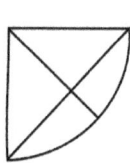 B 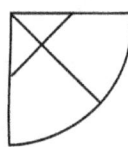 C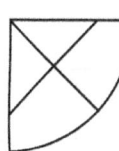

2 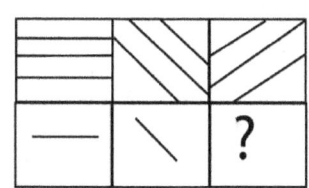 A 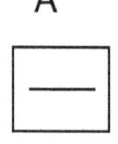 B 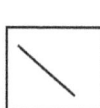 C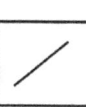

3 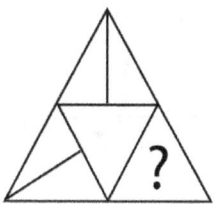 A  B  C

4 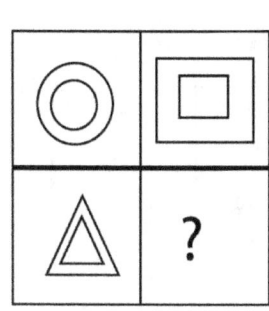 A 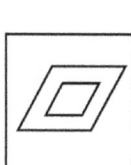 B 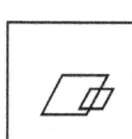 C

Pictorial Reasoning   Answers-154   75
© Gift Of Logic, Inc * Copying prohibited

Name _____ Date _____

PATTERN PERCEPTION - CONTINUING PATTERN

Find the correct figure from the two alternatives given that will logically continue the pattern of figures on the left.

5   A   B

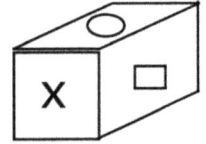   ?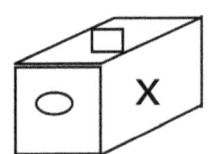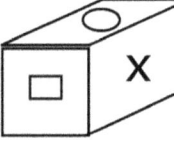

6   A   B

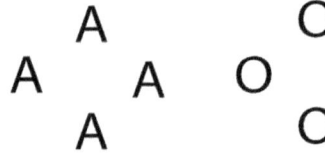   ?   L L   V V
                             L L   V V
                                        V

7   A   B

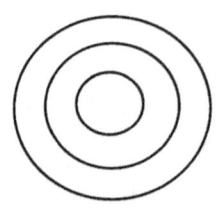   ?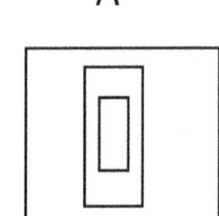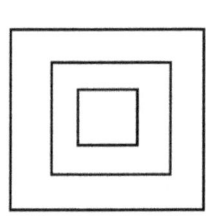

8   A   B

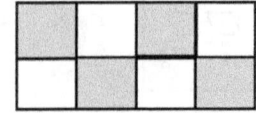   ?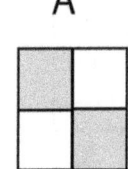

Pictorial Reasoning   Answers-154   76

Name _____   Date _____

FIGURE FORMATION

Find the correct figure that will be formed when the two figures on the left are combined.

1   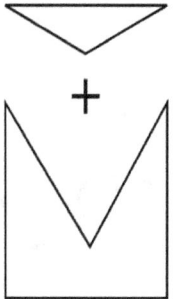   =   A 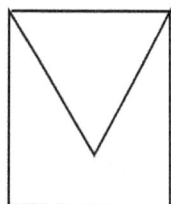   B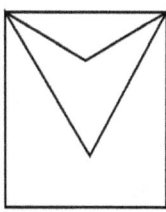

2   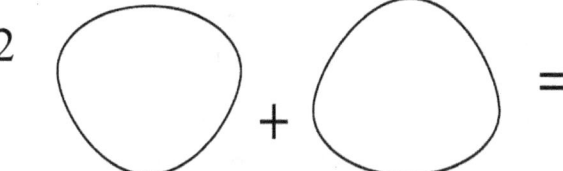   A 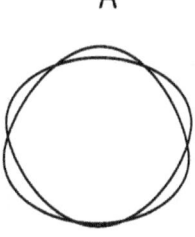   B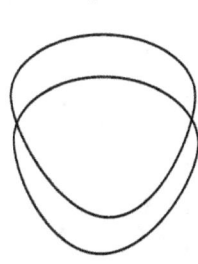

3      =   A 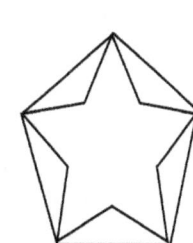   B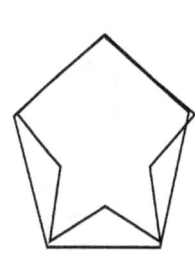

4   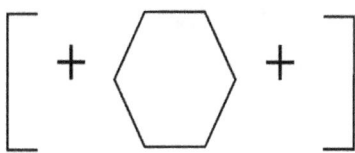   =   A 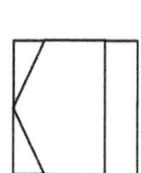   B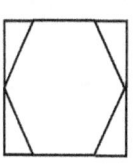

Pictorial Reasoning   Answers-154
© Gift Of Logic, Inc * Copying prohibited

FIGURE FORMATION

Find the correct figure that will be formed when the two figures on the left are combined. Either of the figures may be rotated before combining.

5 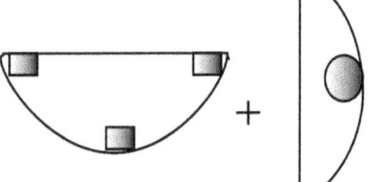 A   B
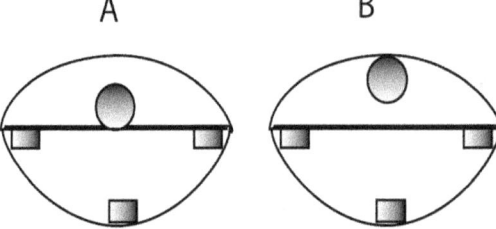

6 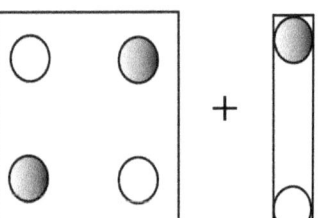   A 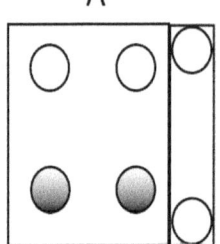   B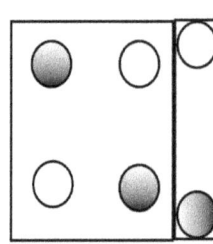

7 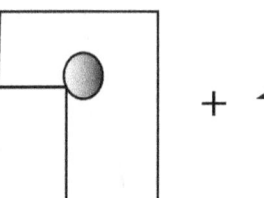   A 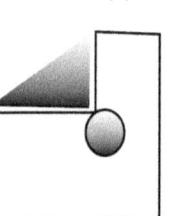   B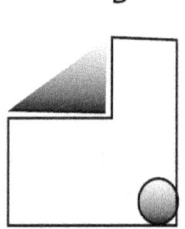

8 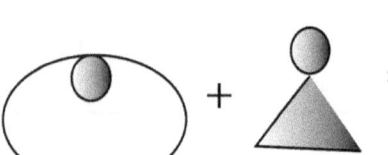   A 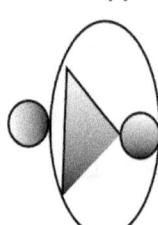   B

Pictorial Reasoning

Name _____   Date _____

PAPER FOLDING AND CUTTING

Find the correct figure that will be formed when the paper on the left is folded in the direction of the arrows, and then holes are cut in it as shown.

1

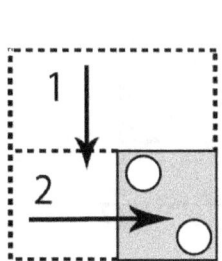 A 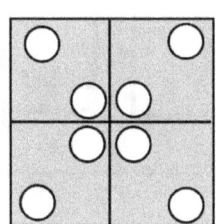 B 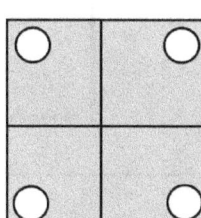 C 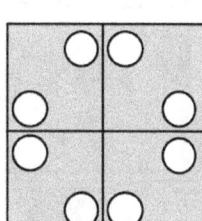 D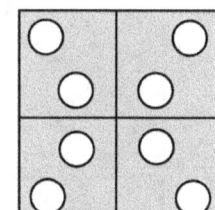

2

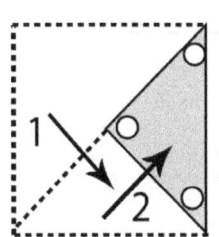 A 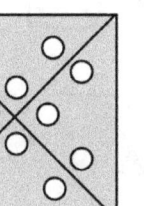 B 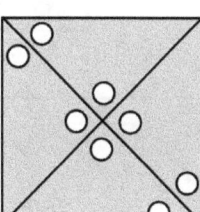 C 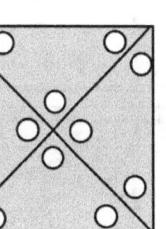 D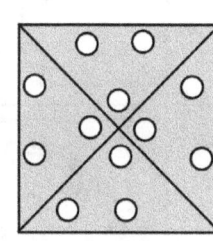

3

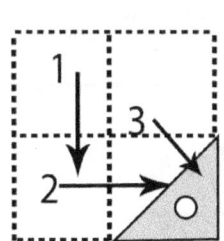 A 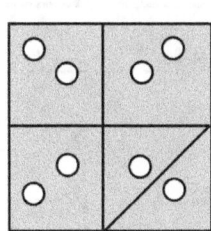 B 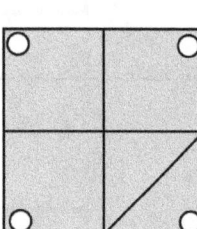 C 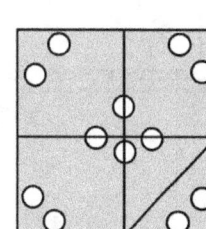 D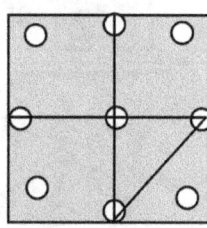

FIGURE MATRIX - ANALOGY

Find the correct figure from the alternatives given that will fit in the empty box such that, the bottom two figures are related in the same way as the top two figures.

1 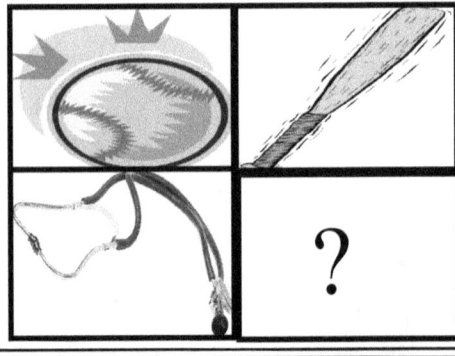 A 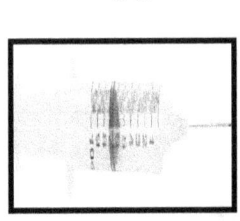 B

2  A 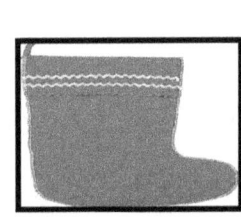 B

3 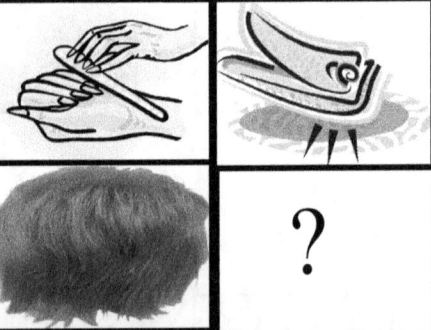 A  B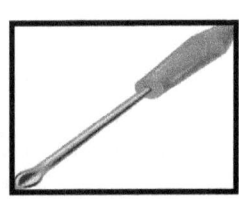

4  A  B

Name _____ Date _____

FIGURE MATRIX - SIMILARITY

Three figures in the 2 x 2 matrix have similar characteristics. Find the fourth figure from the alternatives given that is also alike.

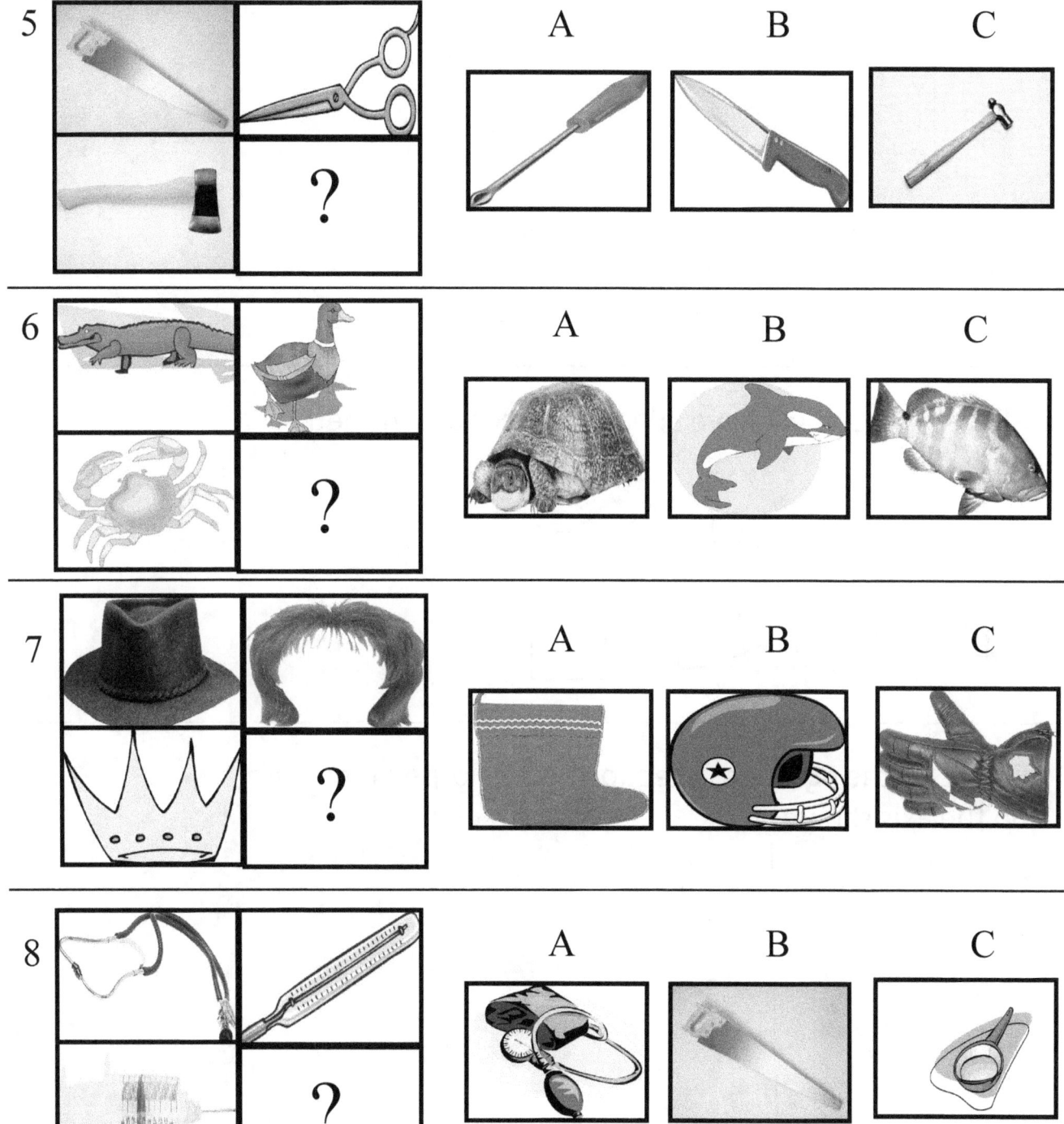

Pictorial Reasoning

RULE DETECTION

Read the given rule in each question. Then, find the correct choice from the alternatives given that satisfies the rule.

1. All the figures must have a vertical line of symmetry

   A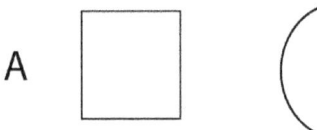

   B

All the figures must have atleast one axis of symmetry

2. A

   B

Figure on the right is a reflection of the figure on the left on the axis shown

3. A

   B

   C

Pictorial Reasoning

Name _____   Date _____

RULE DETECTION

Read the given rule in each question. Then, find the correct choice from the alternatives given that satisfies the rule.

4   The figures must rotate by 90 degrees clockwise

A

B

The figures must rotate by 90 degrees anti-clockwise

5

A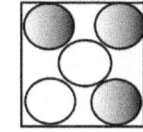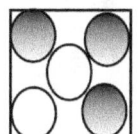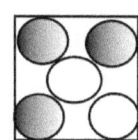

B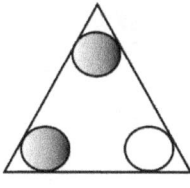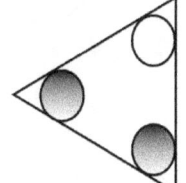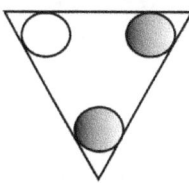

6   The figures must rotate by 45 degrees anti-clockwise

A   ↓   ↘   →   ↗

B   L   <   ⌐   ∧

ANSWERS

1	CONTROVERSY

Dad: Stuart likes Skating, Piano..
Mom: I know that Stuart..

Which one the following represents the point at issue?
A) whether Stuart should take swimming lessons.
B) how many activities Stuart can participate in.
C) whether Stuart should finish his homework in school itself so he can participate in three activities.

ANSWER

Answer: B

Note the conclusions in the two arguments.

Dad's conclusion: Stuart should take lessons in at least three activities.
Mom's conclusion: Stuart can take lessons in at most two activities.

Comparing the conclusions, we can see that the point at issue is how many extracurricular activities Stuart can take part in.

A - incorrect - whether Stuart should take swimming lessons or not is not the point at issue.

B - correct - this is the point at issue. Dad says "at least three" and mom says "at most two". They disagree on this issue.

C - incorrect - both of them do not discuss Stuart doing homework in school itself.

Answers
© Gift Of Logic, Inc * Copying prohibited

2 CONTROVERSY

Cat: People cuddle me..
Dog: It is true that they cuddle..

The controversy between the cat and the dog is regarding which one of the following?
A) who is cuddled more number of times every day.
B) who is loved more.
C) who is cuddled for longer duration.

ANSWER

Answer: B

Cat's conclusion: They love me more than they love you.
Dog's conclusion: Just because they cuddle you more number of times every day does not mean that they love you more.
So, the controversy is regarding who is loved more.

A - incorrect – the cat and dog don't have a disagreement on who is cuddled more number of times everyday. The dog clearly agrees that it gets cuddled only three times a day as opposed to five times a day for the cat.

B - correct - this is what the cat and the dog are arguing about.

C - incorrect – how long it is cuddled is not mentioned by the cat. So they are not arguing about who is cuddled for a longer duration.

3 CONTROVERSY

Josh: I like riding my bike to..
Dave: I like walking..

Which one of the following is the point at issue between Josh and Dave?
A) whether biking to school is better than walking to school.
B) whether everyone should bike to school or walk to school.

ANSWER

Answer: B

Josh's conclusion: Everyone must ride their bike to school.
Dave's conclusion: Everyone must walk to school.

Both arguments take one person's choice and generalizes it for everyone.

A - incorrect – both Dave and Josh do not present facts that compare biking and walking and do not conclude whether one is better than the other. They just recommend biking and walking because they like it.

B - correct – Josh recommends biking to school while Dave recommends walking to school.

Answers
© Gift Of Logic, Inc * Copying prohibited

4 CONTROVERSY

Maggie: This cake will taste..
Jackie: My mom too baked..

What is the point at issue in the debate described above?
A) whether a cake must be baked with lot of sugar or not.
B) whether a layer of vanilla cream is needed or not.
C) whether a cake baked with lot of sugar and topped with vanilla cream will taste good or not.

ANSWER

Answer: C

These arguments do not have conclusion indicators such as "therefore" or "so" and hence the conclusion is a bit difficult to find. But, if you can identify the premises, then you can identify the conclusion.

Maggie's conclusion is "this cake will taste very good". Jackie's conclusion is "Baking a cake with lot of sugar does not guarantee that the cake will taste very good". Both of them present the same premises to reach their conclusion. The premises are "it was baked with a lot of sugar" and "it has a layer of vanilla cream".

A - incorrect – they do not disagree on this issue.

B - incorrect – they do not disagree on this issue

C - correct – this is the point of disagreement

Answers
© Gift Of Logic, Inc * Copying prohibited

5 CONTROVERSY

Customer: The brakes in my car..
Car Dealer: Your car seems to..

What is the point at issue here in the argument between the customer and the car dealer?
A) whether the brakes are working properly or not.
B) whether the problem of the customer's car is unique or not.
C) whether the car dealer should fix the brakes for free or not.

ANSWER

Answer: C

Customer's conclusion: You must fix it for free.
Car Dealer's conclusion: I have to charge you to fix this problem.

A - incorrect – both of them agree that there is a problem with the brakes, but this is not the point of controversy.

B - incorrect – the customer does not say whether the problem is unique to her car. Only the car dealer talks about this. This is not the point at issue.

C - correct – customer wants to fix it for free whereas the car dealer says that he must charge to fix the problem. So, this is the point at issue/controversy.

6   CONTROVERSY

Manager: You have been arriving late..
Employee: My team's morale may be low..

Which one of the following is an issue about which the manager and employee are arguing?
A) whether the employee's tardiness is affecting the team's morale.
B) whether the employee's tardiness is affecting the team's progress.
C) whether the employee is tardy or not.

ANSWER

Answer: B

Manager's conclusion: Your tardiness is affecting the progress and morale of your team.
Employee's conclusion: My tardiness affects the morale of the team, but not the progress (as they are completing their tasks before their deadlines).

Manager says that both the progress and morale is affected, but Employee counters by saying that progress is not affected.

A - incorrect – both agree that the team's morale is affected.

B - correct – this is the point at issue. Employee does not agree that the progress of the team is affected.

C - incorrect – both of them agree about the employee's tardiness.

Answers
© Gift Of Logic, Inc * Copying prohibited

7 CONTROVERSY

Rex: Our competitor, Trojan Computers..
Tina: Trojan Computers ..

Rex and Tina are engaged in a controversy about which one of the following?
A) whether a Trojan computer is better than their own.
B) whether Trojan Computers will sell more computers in all regions.
C) whether the price reduction is in Asia only or not.

ANSWER

Answer: B

Rex's conclusion: Trojan computers will sell more computers than us.
Tina's conclusion: Trojan computers will sell more computers only in Asia.

A - incorrect – this is not the main point at issue. Tina and Rex do not compare the quality of the computers.

B - correct - this is the point at issue. See the conclusion above. Rex indicates that Trojan will sell more computers than his company, but Tina counters that this will happen only in Asia.

C - incorrect - this not the main point at issue. Where the price reduction is effective is the premise that leads to the controversy, but not the controversy itself. The controversy is whether Trojan will sell more computers in all regions or only in Asia.

> **8**   CONTROVERSY

Pro-War Activist: The effect of long term conflict..
Peace Activist: Wars are not effective in putting a permanent..

The point at issue in the argument between the peace activist and the pro-war activist is
A) whether wars will create new conflicts or not.
B) whether wars can end long term conflicts permanently or not.
C) whether long term human suffering must be ended or not.

ANSWER

Answer: B
Note the causal relation: long term conflict c⟶ long term suffering

Pro-war activist's conclusion: Wars are necessary to put a permanent end to long term conflicts.

Peace Activist's conclusion: We need to find another way to terminate long term conflicts.

A - incorrect - neither of them talks about new conflicts.

B - correct – pro-war activist says "wars are necessary .." whereas peace activist says "we need to find another way.. ". So, they disagree on whether wars can end long term conflicts or not.

C - incorrect – this is not the main point at issue. Both of them want to end long term human suffering.

9	CONTROVERSY

Mom: Diana, after reviewing your homeworks..
Diana: Mom, please review it again..

Mom and Diana disagree about which one of the following statements?
A) Diana has not completed assignments A, R and Z.
B) Diana has not completed assignments P and M.

ANSWER

Answer: A

The passage is confusing with a lot of letters and agreements and disagreements. Write them out to help you see the disagreement clearly.

Mom: A P R M Z - not completed
Diana: A R Z - completed, but not other homeworks

A - correct - Diana says that she has completed A, R and Z, but her mom says that she has not. So, they disagree over this statement.

B - incorrect – mom says that Diana has not completed homeworks P and M. Diana says that she has not completed homeworks other than A, R and Z, which means that she agrees that she has not completed P and M. So, both agree over this statement.

Answers
© Gift Of Logic, Inc * Copying prohibited

10 CONTROVERSY

Judge: You have committed..
Defendant: I stole only a few dollars..

Which one of the following expresses the point at issue between the defendant and the judge?

A) the amount of punishment given for minor offense.
B) the amount of punishment that must be given to the defendant.
C) whether committing an offense is immoral.

ANSWER

Answer: A

Judge's conclusion: You deserve the maximum punishment.
Defendant's conclusion: I deserve the minimum punishment.

A - incorrect – the judge does not say that the offense committed is minor nor do they argue about the amount of punishment that is given for minor offenses.

B - correct – the disagreement is over whether the defendant should get maximum or minimum punishment.

C - incorrect – neither of them talk about the morality of committing an offense.

Answers

© Gift Of Logic, Inc * Copying prohibited

11 CONTROVERSY

Principal: Kazaki is the best sportsman..
Assistant Principal: Kazuaki also deserves the gold medal..

The point at issue in the argument between the principal and the assistant principal is

A) whether social service must be considered while deciding who should get the gold medal.

B) whether the gold medal must be shared by both Kazaki and Kazuaki.

ANSWER

Answer: B

Principal's conclusion: No one but Kazaki must get the gold medal.
Assistant principal's conclusion: Kazuaki also deserves the gold medal.

A - incorrect - the principal and the assistant principal do not disagree about considering social service efforts for the gold medal. In fact, both mention the social service efforts of their candidates in positive light.

B - correct - this is the point at issue. Principal says that "no one but Kazaki" must get the gold medal thereby indicating that the gold medal should not be shared, whereas the assistant principal says that Kazuaki also deserves the gold medal, thereby suggesting that the gold medal should be shared. The word "also" indicates "sharing".

12	CONTROVERSY

Rachel: All the artwork submitted to this year's art..
Shannon: The acrylic paintings are certainly breathtaking..

Rachel and Shannon disagree about which one of the following?
A) that the acrylic paintings are very impressive.
B) that oil paintings are of good standard.
C) that beginners can learn from the artwork submitted to the festival.

ANSWER

Answer: C

Rachel and Shannon agree that the acrylic paintings were breathtaking, but do not agree that oil paintings and watercolor paintings are of very high standard. Rachel's argument concludes that the artworks are good for beginners to learn from, but Shannon does not think so.

A - incorrect – both agree that the acrylic paintings are impressive.

B - incorrect - both agree that they are of good standard.

C - correct - Rachel says that beginners can learn from these artworks whereas Shannon says that these artworks are not worthy of mimicry/imitation by beginners, thereby implying that it is of no benefit to the beginners.

13 CONTROVERSY

Mom: Donna, as you know..
Donna: I read today in the Journal of Health..

Mom and Donna disagree on which one of the following?
A) whether watching cartoons beyond a limit will cause mental problems.
B) how long one can watch cartoons before experiencing mental problems.

ANSWER

Answer: B

Mom's conclusion: You must stop watching cartoons now.
Donna's conclusion: I can watch cartoons a bit longer.

A - incorrect – both Donna and her mom agree that watching cartoons beyond a certain limit can cause mental problems. Delirium is a state of mental confusion. They only disagree on the exact limit.

B - correct – Mom says Donna has watched cartoons for more than four hours and thinks that Donna's mental health may be affected. Donna argues that she can watch up to six hours before experiencing any mental effects. So, they disagree over how long one can watch cartoons before experiencing mental problems.

14	CONTROVERSY

King A: My soldiers have captured yours..
King B: My soldiers are still fighting yours..

King B disagrees with King A on which of the following?
A) that his chief of army ran away from the battlefield.
B) that the war is already over.
C) that his soldiers have surrendered.

ANSWER

Answer: C

Note that there several alternate conversations between King A and King B. Read them carefully.

King A's conclusion: Your soldiers have surrendered.
King B's conclusion: My soldiers will fight to the end.

A - incorrect – this is not the point at issue.

B - incorrect – this is not the point at issue. Both of them do not say that the war is over.

C - correct – King A says "your soldiers have surrendered" whereas king B says "my soldiers will fight to the end". So, there is a disagreement over whether the soldiers have surrendered or not.

15 CONTROVERSY

Elizabeth: I like electronic versions..
Paul: I don't like electronic versions..
Elizabeth: That's exactly why I like..
Paul: Well, I like printed books..
Elizabeth: Some electronic books..

Based on the conversation above, Elizabeth and Paul disagree on which one of the following?
A) that one does not have to remember to take electronic books with them. B) that highlighting can be done in electronic versions of books.

ANSWER

Answer: B

A - incorrect – both Elizabeth and Paul agree that one need not remember to take the electronic books with them wherever they go.

B - correct – both of them disagree on this. Paul says that electronic books don't let you highlight words in them whereas Elizabeth says that electronic books let you highlight words in them.

Answers
© Gift Of Logic, Inc * Copying prohibited

16 CONTROVERSY

Mayor: Oil is a natural resource..
Governor: Exploring alternative sources..

The mayor and the governor disagree on which of the following?
A) that oil is a scarce resource.
B) that wind and solar energy are not alternative sources of energy.
C) that penalizing industries for using oil is a good idea.

ANSWER

Answer: C

Mayor's conclusion: "..it is a good idea to penalize industries that depend only on oil.
Governor's conclusion: "...we must not do so"

A - incorrect – this is not the point at issue. They both agree that oil is a scarce resource.

B - incorrect – this is not the point at issue. They do not disagree over this.

C - correct - The Mayor wants to penalize industries that depend only on oil, whereas the Governor does not think that it is a good idea.

Answers
© Gift Of Logic, Inc * Copying prohibited

17 CONTROVERSY

Amanda: Purchasing books from the Internet ..
Vivek: Ordering books from the Internet from the comfort..

Vivek and Amanda are in disagreement over which of the following?
A) that books take a lot of time to get delivered when ordered from the Internet.
B) that you have to go to the book store to place the orders from Internet.
C) that purchasing books from the Internet is always safe.

ANSWER

Answer: C

Amanda's conclusion: Purchasing books from the Internet is convenient and safe.
Vivek's conclusion: Ordering books from the Internet is convenient, but not always safe.

A - incorrect – this is not the point at issue.

B - incorrect – this is not the point at issue.

C - correct – Vivek says that ordering books from the Internet is convenient, but not always safe. Amanda says that "purchasing books from the Internet is convenient and safe", thereby implying it is always safe. The point of disagreement is whether it is always safe to purchase books from the Internet.

Answers
© Gift Of Logic, Inc * Copying prohibited

| 18 | CONTROVERSY |

Coach: If you observe the performance of soccer players..
Principal: If you observe the performance of soccer players..

The Principal and Coach disagree on which of the following?
A) whether only those players who are coached play well.
B) whether soccer players must be coached from five years of age.

ANSWER

Answer: B

Coach's conclusion: ".. coaching for soccer should begin at five years".
Principal's conclusion: "..don't have to necessarily start learning soccer from five years of age".

A - incorrect - this is not the point at issue. They are not arguing whether only players who are coached play well.

B - correct - Coach says that coaching must begin at age five whereas the Principal says that coaching can begin at ten years of age without any difference in performance.

Answers
© Gift Of Logic, Inc * Copying prohibited

19 CONTROVERSY

Bus Driver: Sir, your are sitting in a seat..
Passenger: Since, there is no handicapped..
Bus Driver: The rule is that, even..

The bus driver and the passenger disagree on which of the following?
A) that a seat reserved for the handicapped is meant to be occupied by handicapped people.
B) that a seat reserved for the handicapped must only be occupied by the handicapped at all times.
C) that handicapped people can sit in seats not reserved for them.

ANSWER

Answer: B

A - incorrect – both are in agreement on this issue– bus driver says this strongly and the passenger says that he is justified in sitting in a handicapped seat as long as no handicapped person is looking for a seat – which means that the passenger is in agreement that it is meant to be occupied by handicapped people when they are in the bus.

B - correct – passenger says he is justified in sitting in a reserved seat as long as no handicapped person in the bus, whereas the bus driver cites a rule that says that at all times a seat reserved for the handicapped must not be occupied by the non-handicapped.

C - incorrect – there is no mention of where handicapped people can sit. What is mentioned is who can sit in a seat reserved for the handicapped.

Answers
© Gift Of Logic, Inc * Copying prohibited

| 20 | CONTROVERSY |

Customer: The vacuum cleaner that..
Technician: These vacuum cleaners make..

The customer and the vacuum cleaner technician disagree on which of the following?
A) that the vacuum cleaner makes noise when turned on.
B) that the vacuum cleaner is defective.
C) that it does not suck the dust quickly.

ANSWER

Answer: B

A - incorrect – the customer states that it makes a strange noise – the technician also says that there is a distinct noise heard when operating vacuum cleaners of this type – so they are in agreement on this.

B - correct – customer says it is defective and gives reasons. The technician rebuts each claim of the customer that it is defective and counters that it is normal. In effect, he says that the vacuum cleaner is not defective.

C - incorrect – customer says that it does not suck the dust quickly – technician agrees that it is because of the weak motor in the vacuum cleaner.

21 CONTROVERSY

Doctor Stone: The patient must be given..
Doctor Stewart: The latest bulletin from the head..

Doctor Stone and doctor Stewart are in disagreement over which of the following issues?
A) whether the patient must be given a heavy dose of sedatives.
B) whether the patient is agitated or not.

ANSWER

Answer: A

A - correct – Doctor Stone wants to give a heavy dose of sedatives, but Doctor Stewart prefers only a light dose of sedatives – they are in clear disagreement over this.

B - incorrect – both the doctors are in agreement that the patient is in an agitated state.

22 CONTROVERSY

Pharmacist A: The Doctor has prescribed two tablets..
Pharmacist B: As a matter of policy, the medicines must be dispensed..

The pharmacists agree on the following except
A) that the doctor has prescribed two tablets when a syrup of equal strength is available.
B) that medicines can be dispensed in any form as long as their strengths are the same.

ANSWER

Answer: B
Note that the question is stated differently, but essentially asks you to find the statement over which the pharmacists disagree.

A - incorrect – Pharmacist A says that a syrup of equal strength as the tablets is available. Pharmacist B says that there should be a specific reason why the doctor has not prescribed the syrup. So, both are in agreement that the doctor has prescribed two tablets instead of the syrup.

B - correct – pharmacist A concludes that the syrup can be dispensed since it has the same effect as the two tablets combined – thereby implying that medicines can be dispensed in any form as long as they are of the same strength. Pharmacist B says that as a matter of policy the medicines must be dispensed exactly as prescribed regardless of their form – meaning, one form of medicine cannot be substituted by another form of medicine. So, they are not in agreement over how to dispense medications of equal strengths.

Answers

23 CONTROVERSY

Tenant: Painting the inside of the house is not a violation..
Landlord: The lease agreement states that whenever you paint..

The Tenant and the Landlord disagree on which of the following?
A) whether there has been a violation of the lease agreement.
B) whether the outside of the house must be painted whenever the inside of the house is painted.

ANSWER

Answer: A

Tenant's conclusion: Painting the inside of the house is not a violation of the lease agreement.

Landlord's conclusion: You are in violation of the lease agreement.

A - correct – this is the disagreement. Tenant says that he has not violated the lease agreement whereas the Landlord says that the tenant has violated the lease agreement.

B - incorrect – this is not the point at issue. The tenant does not mention anything about painting the outside of the house.

| 1 | RESOLVE THE PARADOX |

Ronaldo can score three goals..

Which one of the following, if true, explains this paradox?
A) In the final game, his team played against the Panthers, a much stronger team.
B) Ronaldo does not like to play against the Panthers since they wear a red color jersey.
C) Ronaldo was not feeling well during the final game.

ANSWER

Answer: C

Paradox: Ronaldo can score three goals in any game, but in the crucial final game against the Panthers he scored only two goals.

A - incorrect – passage says that he can score three goals in any soccer game. This would include the game against Panthers, even if they are a much stronger team. So, this choice does not explain the reason behind his less-than-expected performance in the final game.

B - incorrect – the color of the jersey that the Panthers were wearing does not explain why he could not score three goals against them.

C - correct – Since he was not feeling well, his performance fell short of expectations in the final game. This statement resolves the paradox.

Answers
© Gift Of Logic, Inc * Copying prohibited

2. RESOLVE THE PARADOX

Trucks carrying hazardous materials..

Which one of the following, if true, most helps to explain the paradox in the passage?

A) motorbikes too are not allowed into the city of Greenville, but you can see them everywhere.
B) hazardous trucks often enter the city in order to save thirty minutes of time driving on the detour route.
C) the detour route has been closed due to an accident that occurred a few minutes ago and the traffic police have authorized the hazardous trucks to enter the city limits temporarily.

ANSWER

Answer: C

Paradox: Trucks carrying hazardous materials are not allowed into the City of Greenville. But, several trucks carrying hazardous chemical (benzene) are inside the city.

A - incorrect – the presence of motorbikes, although they are prohibited inside the city, does not help explain the presence of hazardous trucks.
B - incorrect – trucks are not allowed into the city and must take detour around the city. So, to say that trucks enter the city to save time contradicts the facts and does not explain the paradox.
C - correct – this answer choice resolves the paradox by explaining that trucks were allowed to enter the city temporarily because of an accident.

| 3 | RESOLVE THE PARADOX |

Fingerprints are the patterns..

Which one of the following, if true, explains the puzzle about Travis's fingerprint?
A) Travis is a very religious person and hence would definitely not have stolen from the store.
B) Travis wore gloves while stealing to avoid leaving any fingerprints.
C) Travis had not touched anything when the alarm went off and he left without leaving any fingerprints.

ANSWER

Answer: B

Paradox: A picture of Travis stealing with his hands was captured by a video camera. But, the police could not find any fingerprints of him anywhere in the store.

A- incorrect – this choice appeals to emotion and has no bearing on why fingerprints were not found in the store.

B - correct – by wearing gloves, Travis did not leave any fingerprints, but the video camera recorded pictures of him stealing. This explains the puzzle of the missing fingerprints.

C - incorrect – passage says that video camera clearly shows Travis stealing with his own hands. This choice says that he did not touch anything and hence does not explain the paradox.

Answers

© Gift Of Logic, Inc * Copying prohibited

4 RESOLVE THE PARADOX

Sophia bought a very..

Which one of the following best explains the paradox described above?

A) Sophia is not a professional photographer and so it does not matter whether she uses a cheap camera or an expensive camera.
B) Sophia has the misguided perception that an expensive camera will yield better pictures than an inexpensive one.
C) The new camera was not configured to take advantage of its advanced technology.

ANSWER

Answer: C

Paradox: Sophia's expensive camera has advanced technology that produces crystal clear pictures. But, when she developed the pictures taken with it, they were of poor quality.

A - incorrect – this answer choice, instead of explaining why Sophia was not able to take good pictures using an expensive camera criticizes her credentials as a photographer. This criticism does not explain clearly why the quality of pictures taken with the expensive camera was not good.

B - incorrect – since expensive cameras tend to have more features than inexpensive ones, it is reasonable to expect better quality pictures from it.

C - correct – this choice best explains why Sophia was not able to take better pictures. Had she configured it properly, she would have obtained good quality pictures.

5   RESOLVE THE PARADOX

Hundreds of volunteers and police..

Which one of the following, if true, will explain the widespread unemployment in the city of Oceana?
A) The volunteers did not help rebuild the businesses in the city.
B) The nearby city of Sienna was also hit by the tornado and they too have widespread unemployment.
C) Unemployment is not only in the city of Oceana, it is prevalent everywhere in the country.

ANSWER

Answer: A

Paradox: In spite of the help received from volunteers to rebuild homes, roads, and bridges, there is widespread unemployment in the city of Oceana.

A - correct – this explains why there is widespread unemployment. While the volunteers helped rebuild homes, roads and bridges, they did not help rebuild the businesses.

B - incorrect – the question asks to explain the unemployment in the city of Oceana – the answer choice talks about the city of Sienna which is not relevant.

C - incorrect – the fact that there is unemployment everywhere in the country does not explain why in spite of building homes, roads and bridges, the city of Oceana still has widespread unemployment.

Answers
© Gift Of Logic, Inc * Copying prohibited

6 RESOLVE THE PARADOX

Farmer Joe purchased tomato seeds..

Which one of the following, if true, most helps to resolve the mystery behind the incorrect size of the tomatoes?

A) Farmer Joe should have purchased the seeds from Evergreen Nursery which carries better quality seeds than Greenhouse Nursery.
B) Other farmers bought the exact same seeds and obtained large, red and delicious tomatoes.
C) Farmer Joe did not apply the exact amount of fertilizer that was recommended by Greenhouse Nursery for growing large tomatoes.

ANSWER

Answer: C

Paradox: Farmer Joe expected large tomatoes, but was disappointed when he found only small tomatoes.

A - incorrect – what the Evergreen nursery sells has no bearing on why farmer Joe's tomato plants yielded only small sized tomatoes.

B - incorrect - the size of fruits other farmers got from their plants does not explain why farmer Joe got small sized tomatoes.

C - correct – because he did not apply the recommended amount of fertilizer, he did not get the large tomatoes. This choice resolves the paradox.

Answers

| 7 | RESOLVE THE PARADOX |

A fitness club has two types of memberships..

Which one of the following, if true, would help to resolve the puzzle experienced by the owner of the fitness club?
A) The fitness club conducts several free fitness programs specially for children.
B) The five feet deep swimming pool is unsuitable for most children.
C) Most people who are eligible for the "Single" membership do not like to go to a fitness club that has family memberships.

ANSWER

Answer: B

Puzzle: There are more families in the neighborhood than single people. But, a majority of the club's customers have single membership.

A - incorrect – if it conducted several free programs to benefit the children, then the family membership would be high, but it is not. So, this does not explain the puzzle.

B - correct – since the swimming pool is not suitable for most children, parents will not bring their children to the club. This will explain its low "family" membership levels.

C - incorrect – if this were the case, then adults would not come to the fitness club. But, the passage says that majority of customers do not have the "family" membership, which means that they have the "Single" membership. This indicates that adults do come to the club.

Answers
© Gift Of Logic, Inc * Copying prohibited

8 RESOLVE THE PARADOX

Traffic in the Interstate Highway #75..

Which one of the following, if true, will help resolve the paradox described above?

A) People driving in the freeway after peak hours take only twenty minutes to travel ten miles in the freeway.
B) New commuters have started using the freeway to take advantage of the additional lane.
C) The speed limit during peak hours was increased to relieve congestion.

ANSWER

Answer: B

Paradox: An additional lane was opened to relieve congestion during peak hours. But, the time to travel ten miles has increased more instead of decreasing.

A - incorrect – read carefully - the passage is about the woes of peak hour commuters, but not after peak hour commuters. The additional lane was opened to alleviate the problem for peak hour commuters.

B - correct – if new commuters got attracted to the freeway because of the additional lane, it would offset any benefit that would come from the additional lane. This would explain why, in spite of the additional lane, the travel time has become worse.

C - incorrect – if the speed limit was increased, it would have decreased the travel time and not make things worse. So, this does not explain the extra delay due to the opening up of the new lane.

Answers

9 RESOLVE THE PARADOX

Rhonda saw a doctor to treat a small skin infection..

Which one of the following, if true, will help to explain Rhonda's disappointment?
A) Rhonda used the ointment on her legs last year and the itching caused by the skin infection went away within two days.
B) The ointment created an adverse reaction when it interacted with the metallic ring in her fingers.
C) Rhonda washed her hands three times a day to help prevent skin infections.

ANSWER

Answer: B
Paradox: Rhonda used an ointment that was expected to provide relief to itching within two weeks, but even after two weeks, her itching has not gone away.

A - incorrect – the passage talks about a skin infection in Rhonda's hands. Her use of the ointment in her legs does not explain why the itching in her hand still remains.

B - correct – since the ointment created an adverse reaction due to its interaction with the metallic ring in her fingers, this choice explains why her infection has not gone away even after two weeks.

C - incorrect- if she washed her hands three times a day, with the hope that it would prevent skin infections, her current infection should have healed as well, but it has not. So, this does not explain why the infection still remains.

Answers
© Gift Of Logic, Inc * Copying prohibited

10 RESOLVE THE PARADOX

Molly's mom hired a private tutor..

Which one of the following, if true, would help resolve the paradox?

A) Molly's tutor taught her more than what was necessary.
B) Molly did not take the tutorials seriously.
C) Students with private tutors show a significant improvement in their academic performance.

ANSWER

Answer: B

Paradox: A tutor was hired to help Molly with her academic performance. But, in spite of this, her academic performance continued to decline.

A - incorrect - if Molly's tutor taught her more than what was necessary, it is unlikely to have caused a decline in her performance.

B - correct- since she did not take advantage of the tutorials seriously, she did not benefit from it and hence her continued decline in academic performance.

C - incorrect – if this were the case, Molly's performance would have improved, but in fact it only declined.

| 11 | RESOLVE THE PARADOX |

The Creek Hollow community of houses ..

Which one of the following, if true, most helps to explain the paradox described above?
A) The home owners have decorated the entrances to Creek Hollow with pretty flowers.
B) Lot of young couples have purchased homes in the community.
C) Due to lack of jobs in the city where the Creek Hollow community is located, the number of home buyers have significantly reduced.

ANSWER

Answer: C

Paradox: Houses in the community used to sell out quickly. Surprisingly, it now takes twice as much time to sell them.

A - incorrect – decorating the entrances to the community with pretty flowers would only help in adding value to the home and speed up the selling – not delay the selling time.

B - incorrect - this does not explain why the houses are taking more time to sell.

C - correct – since the number of home buyers have reduced, it takes longer to sell it.

12 RESOLVE THE PARADOX

The race for the Miss Galaxy award..

Which one of the following, if true, best explains the reason for this puzzling decision regarding the recipient of the "Miss Galaxy" award?

A) Internet is accessible to a majority of people in Anita's country.
B) People in Nicole's country have the most access to internet.
C) The viewer's internet-vote was the only factor that decided the winner.
D) The viewer's internet-vote played only a minor role in the selection of the winner.

ANSWER

Answer: D

Paradox: Nicole was selected by internet-voting as the "Miss Galaxy", but Anita was crowned as "Miss Galaxy" instead.

A - incorrect – even though internet is available to a majority of people in Anita's country, it was Nicole who was selected by internet-voting. So, this choice does not explain why Anita was crowned as "Miss Galaxy".

B - incorrect – this choice does not explain why she was not crowned "Miss Galaxy".

C - incorrect – if this were the case, then Nicole would have won, but she did not. So, this does not explain the puzzle.

| 13 | RESOLVE THE PARADOX |

A 24-theater mega movie complex..

Which one of the following, if true, most helps to explain the low ticket sales in the movie theater complex?
A) There is a trash-collecting facility near the theater complex.
B) The employees of the theater complex do not like Mr. Strongman.
C) There are already several well established movie theater complexes in the city.

ANSWER

Answer: C

Paradox: High ticket sales was expected from the new movie theater complex, but the number of visitors to the complex is much lower than estimated.

A - incorrect – since it is normal to have a trash collecting facility in an area visited by a large number of people, this choice does not explain why the ticket sales are low.

B - incorrect – Mr. Strongman's personality has no bearing on ticket sales. Visitors to the complex may not even know that he is the owner or interact with him.

C - correct – since the city already has several well established movie theater complexes, addition of a new movie theater complex did not yield Mr. Strongman the ticket sales that he expected.

Answers
© Gift Of Logic, Inc * Copying prohibited

14 RESOLVE THE PARADOX

Dad: There were ten candies..
Alex: Honestly, I did leave two candies..

Which one of the following, if true, most helps to reconcile Alex's response to his Dad's question?
A) After eating eight candies, Alex put the remaining two in the box.
B) There were no holes in the box that held the candies.
C) Alex did not notice that the eighth and ninth candies were attached to each other.

ANSWER

Answer: C

Paradox: There were supposed to be two candies in the box, but there is only one candy.

A - incorrect – if he did put the remaining two candies in the box, that does not explain why there is only one remaining.

B - incorrect – this does not explain why one candy is missing.

C - correct – this means that when he wrote 8 in his hand, he actually ate the eighth and the ninth candy, a total of nine candies. This explains why only one candy was left in the box for his brother.

Answers

15 RESOLVE THE PARADOX

Home-Owner: A very intense fire started..
Firefighter: Yes, I can see that the fire was very intense indeed..

Which one of the following, if true, most helps to explain why there was very little damage to the house?
A) The intensity of the fire caused expensive damage to the bedroom and the adjacent living room.
B) Metal fragments from the burned out room heater were found to be at a temperature of 500 degrees Fahrenheit.
C) The intensity of the fire turned on an automatic sprinkler system that doused the flames.

ANSWER

Answer: C

Paradox: A very intense fire was found in the bedroom, but very little damage was done.

A - incorrect – this is contradictory to the firefighter's opinion that there is very little damage to the house.

B - incorrect – this only supports what is stated by the home owner and the firefighter – that there was an intense fire – it does not explain why it did not cause much damage.

C - correct – the fact that the fire was doused by water from an automatic sprinkler system explains why there was very little damage to the house.

16 RESOLVE THE PARADOX

Several families with school-going children..

Which one of the following, if true, most helps to explain how the quality of education did not deteriorate because of the additional student load?

A) Seven new teachers were hired to help the current teachers.
B) The student teacher ratio was maintained at the same level after the hurricane as it was before the hurricane.

ANSWER

Answer: B

Paradox: Two hundred extra students were admitted to Kindness Elementary School, but surprisingly, the quality of education received by the students was not affected.

A - incorrect - this could be the correct answer, but it does not clearly explain whether the seven new teachers were sufficient for handling the extra student load.

B - correct - this choice explains that the same student teacher ratio was maintained after the addition of two hundred students. This means that sufficient number of additional teachers were hired to handle the extra student load. This choice explains the paradox.

Answers

17 RESOLVE THE PARADOX

Four new 60 Watt bulbs, all made..

Which one of the following, if true, will resolve the paradox presented above?

A) Just like human beings born on the same day don't die on the same day, bulbs too have different life spans.
B) It is not realistic to expect four new, exactly same type of bulbs to have the same longevity.
C) The vacuum-creating process, a key factor deciding the longevity of bulbs, is not a perfect process.

ANSWER

Paradox: Four new bulbs made by the same company were turned on at the same time. But, they did not burn out at the same time.

Answer: C

A - incorrect - comparing longevity of human beings to longevity of bulbs is not meaningful and does not explain the paradox.

B - incorrect - this choice is more of an opinion than an explanation. Some may disagree with this opinion.

C - correct – if the vacuum-creating process, a key factor deciding the longevity of the bulbs, is not perfect, then four new bulbs would have inherent differences when they were manufactured. So, this would explain why they don't have the same longevity and burned out at different times.

1 SUDOKU

Solve the following Sudoku. A correctly solved Sudoku has numbers 1-9 appearing only once in each row, each column and each 3x3 grid. Solving Sudokus will help you to gain valuable analytic skills.

3	5	4	6	9	1	8	2	7
8	6	2	4	5	7	9	3	1
1	9	7	3	8	2	4	5	6
6	1	3	2	7	4	5	9	8
7	8	9	5	1	3	2	6	4
2	4	5	9	6	8	7	1	3
9	2	1	7	4	6	3	8	5
4	3	8	1	2	5	6	7	9
5	7	6	8	3	9	1	4	2

Answers

2 SUDOKU

Solve the following Sudoku. A correctly solved Sudoku has numbers 1-9 appearing only once in each row, each column and each 3x3 grid. Solving Sudokus will help you to gain valuable analytic skills.

1	3	8	7	9	5	6	4	2
4	7	9	6	3	2	1	5	8
2	6	5	1	4	8	3	7	9
3	5	1	8	7	9	4	2	6
7	9	6	2	1	4	5	8	3
8	4	2	5	6	3	7	9	1
6	8	3	4	2	7	9	1	5
5	1	7	9	8	6	2	3	4
9	2	4	3	5	1	8	6	7

Answers
© Gift Of Logic, Inc * Copying prohibited

3

SUDOKU

Solve the following Sudoku. A correctly solved Sudoku has numbers 1-9 appearing only once in each row, each column and each 3x3 grid. Solving Sudokus will help you to gain valuable analytic skills.

8	3	5	1	2	9	7	4	6
9	7	1	4	6	5	2	3	8
2	6	4	3	7	8	1	5	9
3	1	9	2	8	6	5	7	4
4	8	2	5	1	7	6	9	3
7	5	6	9	3	4	8	2	1
5	4	8	6	9	2	3	1	7
6	2	3	7	4	1	9	8	5
1	9	7	8	5	3	4	6	2

Answers

© Gift Of Logic, Inc * Copying prohibited

4
SUDOKU

Solve the following Sudoku. A correctly solved Sudoku has numbers 1-9 appearing only once in each row, each column and each 3x3 grid. Solving Sudokus will help you to gain valuable analytic skills.

9	7	3	8	6	1	5	2	4
1	8	4	3	2	5	6	9	7
5	2	6	7	4	9	1	8	3
7	9	5	6	3	8	2	4	1
3	1	8	2	9	4	7	6	5
4	6	2	5	1	7	8	3	9
6	3	1	4	7	2	9	5	8
2	5	9	1	8	3	4	7	6
8	4	7	9	5	6	3	1	2

Answers
© Gift Of Logic, Inc * Copying prohibited

5 SUDOKU

Solve the following Sudoku. A correctly solved Sudoku has numbers 1-9 appearing only once in each row, each column and each 3x3 grid. Solving Sudokus will help you to gain valuable analytic skills.

1	7	2	8	4	5	6	3	9
3	4	9	7	2	6	5	8	1
6	5	8	9	3	1	2	4	7
2	1	4	6	8	9	3	7	5
7	8	6	2	5	3	1	9	4
5	9	3	4	1	7	8	2	6
4	3	1	5	9	8	7	6	2
9	6	5	3	7	2	4	1	8
8	2	7	1	6	4	9	5	3

Answers

POSITIONING

1

Two boys and two girls..

1) How many seating configurations are possible? Write them below.

Six seating configurations are possible as shown below.
    Boy, Boy, Girl, Girl
    Boy, Girl, Boy, Girl
    Girl, Girl, Boy, Boy
    Girl, Boy, Girl, Boy
    Girl, Boy, Boy, Girl
    Boy, Girl, Girl, Boy

2

Two boys and two girls..
A boy and a girl must sit next to each other.

1) If a boy sits in the fourth chair, another boy can sit in the first chair.
    A) True   B) False

Answer: B) False. If a boy sits in the fourth chair, then according to the rule, the seating will be: Girl, Boy, Girl, Boy. So, another boy cannot sit in the first chair.

Answers
© Gift Of Logic, Inc * Copying prohibited

POSITIONING

3 Place the following animals ..

Diagram the scenario first using symbols.

Zebra, Horse, Elephant, Monkey >> Z,H,E,M
Zebra must be in the first spot >> Z1
Monkey must be next to the Zebra. >> ZM (since Z is in the first spot, MZ is not possible as there is no spot to the left of spot# 1)
Horse must be to the right of the Elephant.>> E-H

Now, place the animals in the spots.

Z,H,E,M
Z1
ZM
E-H

1	2	3	4
Z	M		

After placing Z in spot# 1 and M in spot# 2 (to satisfy rule ZM), we are left with positions 3 and 4. Elephant and Horse has to be placed in spots 3 and 4 according to the rule E-H, Since there are only 2 spots available, there cannot be anyone between E and H. So, E has to be in spot# 3 and H has to be in spot# 4. The final positions are shown below.

Z,H,E,M
Z1
ZM
E-H

1	2	3	4
Z	M	E	H

Answers
© Gift Of Logic, Inc * Copying prohibited

POSITIONING

4 Diagram the scenario using symbols.

Three birds, a Parrot, a Peacock, and an Eagle must be placed in three cages shown according to the rules described below. >> Pa, Pe, E

Parrot must be immediately to the left of Peacock. >> PaPe
Peacock must be in the third cage. >> Pe3

Now, place the birds in the cages according to the rules. Pe (peacock) is placed in the third cage, Pa(parrot) is placed in the second cage (to satisfy PaPe). Now eagle has no choice, but to be in cage# 1.

Pa,Pe,E
Pe3
PaPe

1	2	3
E	Pa	Pe

1) In which cage can the Parrot be placed?
Answer: B) cage# 2. From the diagram, it is clear that since Peacock has to be in the third cage and Parrot has to be immediately to the left of the Peacock, cage# 2 is the only spot that Parrot can take.

2) In which cage can the Eagle be placed?
Answer: B) cage# 1. This is clear from the diagram shown above.

Answers
© Gift Of Logic, Inc * Copying prohibited

POSITIONING

5 Four balls, yellow, red, blue, and green in color must be placed in the boxes. >> Y,R,B,G (Yellow, Red, Blue and Green)

Diagram the scenario as follows.

Yellow ball must be in box# 1. >>Y1
Blue ball must be to the right of the red ball.>> R-B

Y,R,B,G
Y1
R-B

1	2	3	4
Y			

1) Can the Blue ball be placed in box# 2?

Y,R,B,G
Y1
R-B

1	2	3	4
Y	B		

Answer: B) No. Try to place the blue ball in box# 2. You will immediately notice that this is not possible as it violates the R-B rule. The Red ball cannot be placed before the blue ball since there is only one box before box# 2 and it is taken by the Yellow ball. You can appreciate the fact that diagramming helps you to eliminate impossible positions and verify the correct positions quickly.

Answers
© Gift Of Logic, Inc * Copying prohibited

POSITIONING

2) If the Blue ball is placed in box# 3, where can the Red ball be placed?

Y,R,B,G
Y1
R-B

1	2	3	4
Y	R	B	G

Answer: A) box# 2. From the diagram, which shows B in box# 3, we can see that to satisfy the R-B rule, R has to be in box# 2 only.

3) If the Blue ball is in box# 4, where can the Red ball be placed?

Y,R,B,G
Y1
R-B

1	2	3	4
Y	R	G	B
Y	G	R	B

Answer: C) box# 2 or box# 3. If the Blue ball takes box# 4, since the Yellow ball is in box# 1, Red can take either box# 2 or box# 3 and still satisfy the R-B rule. The two possibilities are shown in the two rows.

Answers
© Gift Of Logic, Inc * Copying prohibited

POSITIONING

6 Four plants are to be planted in four spots shown above..

Note that the rules pertain to Rose and Jasmine, but this problem does not say how many plants are Rose plants and how many are Jasmine plants.

Jasmine must be planted in spot 1 or spot 3, but nowhere else. >> J1|| J3. This rule does not say "but not both". So, Jasmine can be planted in spots 1 or 3 or both. This is a case of inclusive OR.

Rose must be planted in spot 2 or spot 4 only. >> R2 || R4. This is also a case of inclusive OR.

1) Which of the following plantings is correct?

J,R
J1||J3
R2||R4

1	2	3	4
J	R	R	J
J	R	J	R

Answer B) spot-1: jasmine spot-2: rose spot-3: jasmine spot-4: rose

The two answer choices are placed in two rows shown above. Choice A is incorrect because J is planted in spot# 4. This violates J1|| J3.

Choice B is correct since it satisfies the rules.

POSITIONING

7 Three spots are available for planting..
Either carrot or tomato can be planted in spot# 1, but not both.
   C1 ⊮ T1
Either tomato or peach can be planted in spot# 2, but not both.
   T2 ⊮ P2
Either peach or carrot can be planted in spot# 3, but not both.
   P3 ⊮ C3
Note the "but not both" in the rules. This is shown as ⊮ (exclusive or).

1) If tomato is planted in two spots, then carrot cannot be planted.

C1 ⊮ T1
T2 ⊮ P2
P3 ⊮ C3

1	2	3
T	T	C

Answer: B) False. If tomato is planted in two spots, it can only be possible in spots 1 and 2. In spot# 3, carrot can be planted per rule P3 ⊮ C3. So, the answer is False.

2) If tomato is planted in spot# 2 and carrot is planted in spot# 3, then peach cannot be planted. Answer: A) True. Since spots 2 and 3 are taken, we are left with spot# 1 only. In spot# 1, either carrot or tomato can be planted. So, peach cannot be planted.

C1 ⊮ T1
T2 ⊮ P2
P3 ⊮ C3

1	2	3
	T	C

Answers
© Gift Of Logic, Inc * Copying prohibited

POSITIONING

8

Boxes 1 and 3 must be painted red - R1, R3
Box# 2 can be painted green or blue. G2 ∦ B2 (∦ symbol for exclusive OR is used since only one color can be used to paint a box)
Box# 4 can be blue or yellow. B4 ∦ Y4

1) Which of the following choice of colors is valid?

Mark each choice in the grid and compare them with the symbolic rules and easily answer the question.

R1,R3
G2 ∦ B2
B4 ∦ Y4

1	2	3	4
R	B	G	B
R	G	R	Y
R	B	R	G
R	B	R	B

A) red   blue   green   blue >> violates R3
B) red   green red   yellow >> correct
C) red   blue   red   green >> violates B4 ∦ Y4
D) red   blue   red   blue   >> correct

Answers

POSITIONING

9

The rule" The boxes at the ends must be painted red and green", does not say which end must be red and which end must be green. This rule effectively means that if we paint red at one end, we must paint green at the other end and vice versa. We can represent this rule as R1G4∦ R4G1.

Represent the rule "Box# 2 must be painted blue and box# 3 must be painted yellow" with symbols B2, Y3

Q) Which of the following choice of colors is correct for the boxes?

B2,Y3
R1G4 ∦ R4G1

1	2	3	4
R	B	Y	G
G	B	Y	R
R	B	G	Y
G	R	B	Y

Place the choices in the diagram and see if any rule is violated.

A) red, blue, yellow, green >> correct
B) green, blue, yellow, red >> correct
C) red, blue, green, yellow >> incorrect, violates R1G4
D) green, red, blue, yellow >> incorrect, violates R4G1

Answers

© Gift Of Logic, Inc * Copying prohibited

POSITIONING

10

Gina, Gemma, and Gordon...

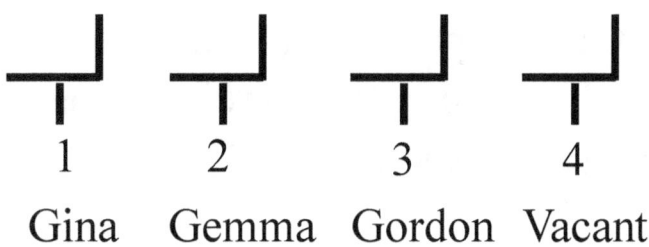

     1   2   3   4

  Gina Gemma Gordon Vacant

1) If Gordon moves to position 4, which chair will become vacant?
Answer: A) 3. This is obvious.

2) If Gemma and Gordon each move one position to the right, which chair will become vacant?

Answer: A) 2. If Gemma and Gordon each move one position to the right, Gordon will be in chair# 4 and Gemma will be in chair #3, leaving chair# 2 vacant.

3) Which of the following moves will make the chair next to Gina vacant?

Answer: B) Move Gemma to chair# 4. The chair next to Gina is chair # 2. This will be vacant after Gemma moves to chair# 4.

Answers
© Gift Of Logic, Inc * Copying prohibited

POSITIONING

11 Nancy, Nina, Nick, and Noor...

First, diagram the rules so that you can answer the questions easily. Since the first letters of all the four persons start with N, using only one letter will be confusing. Use two letters to refer to the persons. Ni-Nina, Nk-Nick, Na-Nancy, No-Noor.

The first rule, Nina and Nick must sit at the chairs at the corners is represented as Ni1,Nk4 ∦ Nk1,Ni4

The second rule that Noor must sit to the left of Nancy is represented as No-Na, the hyphen "-" means that there may or may not be someone between them. Now, we can answer the questions.

1) Nancy must sit in the third chair.

Ni,Nk,Na,No
No-Na
Ni1,Nk4 ∦ Nk1,Ni4

1	2	3	4
Ni		Na	Nk

Answer: A) True. This is the only chair in which Nancy can sit. Nina and Nk have to sit in chairs 1 and 4, which leaves chairs 2 and 3 for Noor and Nancy. But, since Noor has to sit the left of Nancy (No-Na), Nancy can only take chair# 3.

Answers
© Gift Of Logic, Inc * Copying prohibited

POSITIONING

2) Nina must be sit in the first chair.
Answer: B) False. The rule is Ni1,Nk4╫Nk1,Ni4. So, Nina can sit in the first or the fourth chair.

3) Nick must sit in the fourth chair.
Answer: B) False. Again, the rule is Ni1,Nk4 ╫ Nk1,Ni4. This means that Nick can sit either in the first chair or the fourth chair.

4) Nina must always sit to the left of Noor.

Ni,Nk,Na,No
No-Na
Ni1,Nk4 ╫ Nk1,Ni4

1	2	3	4
Nk	No	Na	Ni

Answer: B) False. See diagram above. If the four sit in the order of Nick, Noor, Nancy, Nina then Nina can sit to the right of Noor.

5) Noor cannot sit in the third chair.

Ni,Nk,Na,No
No-Na
Ni1,Nk4 ╫ Nk1,Ni4

1	2	3	4
Nk		No	Ni

Answer: A) True. If Noor sits in the third chair, then the rule No-Na says that Nancy has to sit to the right of Noor in the fourth chair, but this is not possible as the fourth chair is reserved for either Nick or Nina.

Answers
© Gift Of Logic, Inc * Copying prohibited

POSITIONING

12 Harry, Hilga, Hughes, and Heather are to be seated in four consecutive positions. >> Ha,Hi,Hu,He
Harry can be in position 1 or 4 only. >> Ha1∦Ha4
Hilga and Heather must sit next to each other >> HiHe ∦ HeHi

1) If Hughes sits in position 1, Harry can sit in which of the following positions?

Ha1∦Ha4
HiHe ∦ HeHi

1	2	3	4
Hu			Ha

Answer: C) 4. Harry can sit in only 2 positions, 1 or 4. Since Hughes has already taken position 1, Harry has to take position 4.

2) If Harry is in position 1 and Hughes is in position 4, Hilga must be in position 2.

Ha1∦Ha4
HiHe ∦ HeHi

1	2	3	4
Ha	Hi	He	Hu
Ha	He	Hi	Hu

Answer: B) False. If positions 1 and 4 are taken by Harry and Hughes respectively, then positions 2 and 3 will be vacant for Hilga and Heather. The rule states that Hilga and Heather have to sit next to each other. So, if Hilga is in position 2, Heather must be in position 3 and vice versa. So, Hilga can be in position 2 or in position 3.

Answers

© Gift Of Logic, Inc * Copying prohibited

POSITIONING

13 Ian, Irene, Iqbal, and Ishwar ..>> Ia, Ir, Iq, Is
Either Ian or Irene can sit in the first chair. >> Ia1 ⫮ Ir1
Either Irene or Iqbal can sit in the second chair.>> Ir2 ⫮ Iq2
Either Iqbal or Ishwar can sit in the third chair. >> Iq3 ⫮ Is3
Either Ishwar or Ian can sit in the fourth chair. >> Is4 ⫮ Ia4

1) If Ian sits in the first chair, Ishwar can sit in which of the following chairs?

Ia, Ir, Iq, Is
Ia1 ⫮ Ir1
Ir2 ⫮ Iq2
Iq3 ⫮ Is3
Is4 ⫮ Ia4

1	2	3	4
Ia	Ir	Iq	Is

Answer: B) 4. If Ian sits in the first chair, then Irene cannot sit in the first chair. So, Irene has to sit in the second chair, which means that Iqbal cannot sit in the second chair, but can do so in the third chair. If Iqbal sits in the third chair, then Ishwar has to sit in the fourth chair.

2) If Ian sits in the fourth chair, Irene can sit in which of the following chairs?
Answer: A) 1. Using reasoning similar to the answer to the previous question, if Ian sits in chair#4 then Irene can sit only in chair# 1.
3) If Iqbal sits in the third chair, Ian can sit in the fourth chair.
Answer: B) False. If Iqbal sits in the third chair, then Ishwar has to sit in chair# 4. This means that Ian can not sit in the fourth chair. So, the answer to this question is False. Ian will have to sit in the first chair.

Answers

POSITIONING

14 A parrot, a butterfly, an eagle, and a flamingo... >> P,B,E,F
The butterfly must not be next to the eagle.>> ~BE, ~EB
The parrot must not be in the first spot. >> ~P1

1) Which of the following is a correct positioning of the birds?

Answer: B) Butterfly, Parrot, Eagle, Flamingo. This satisfies both the conditions namely ~ P1 and ~BE, ~EB. Answer choice A is incorrect because it does not satisfy the ~P1 rule.

---

15 D-Dog, C-Cat, E-Elephant, L-Lion
The Dog must be in first spot. >> D1
The Elephant must not be next to the dog.>> ~ED, ~DE
The Lion must be to the right of the elephant >> E-L

1) The Cat must be in which spot? Answer: A) 2. Cat can only be in spot# 2.

D1
~ED
~DE
E-L

1	2	3	4
D	C	E	L

2) The Elephant cannot be in spot 4. Answer: A) True
To answer if this assertion is true or false, see what happens if the elephant is placed in spot 4. Then according to rule E-L, Lion must be to the right of the Elephant, but this is not possible. So, the elephant cannot be in spot # 4.

Answers
© Gift Of Logic, Inc * Copying prohibited

POSITIONING

16 Nancy and Olivia..
Nancy must sit before Olivia. >> N-O
Nancy must sit immediately after Mary. >> MN

1) Which of the following seatings is correct based on the above rules?
Answer: B) Mary, Nancy, Pat, Olivia.

From the rules, you need to look for MN and N-O (or MN-O) in the answer choices. This seating arrangement meets the conditions and is the correct answer. Choice A is incorrect because it does not satisfy the N-O rule.

---

17 Roy and Uday ..

Diagram the rules as follows.

Roy cannot stand before Sam. >> ~R-S
Uday cannot stand after Tom. >> ~T-U

1) Which of the following positioning is correct?
   A) Sam, Rohan, Uday, Tom >> correct answer
   B) Sam, Rohan, Tom, Uday >> does not satisfy the ~T-U rule
   C) Rohan, Uday, Sam, Tom >> does not satisfy the ~R-S rule

Answers
© Gift Of Logic, Inc * Copying prohibited

1	GROUPING

Three boys, Ali, Bob, and Chen..

1) How many chess games must be played so that each boy plays the other. Write the names of the players who play in each game.

　　　Game# 1　Ali, Bob
　　　Game# 2　Bob, Chen
　　　Game# 3　Chen, Ali

2	GROUPING

Three girls Asha, Brandi, and Christina went to a movie..

The rule "If Asha watches the movie, then Brandi also must watch the movie" is a condition that can be represented as A → B. Apply this rule to each of the answer choices.

1) Which of the following pairs can watch the movie?

　　A)　Asha, Brandi >> Possible
　　B)　Brandi, Christina >> Possible, A → B does not mean B → A.
　　C)　Christina, Asha >> Not possible, violates the rule A → B.

Answers
© Gift Of Logic, Inc * Copying prohibited

3	**GROUPING**

Three animals, an Anteater, a Bison, and a Cheetah ..

The rule "If and only if the Anteater goes for the walk will the Bison go for the walk" has the biconditional "if and only if" in it. This can be represented as follows: A ↔ B. This biconditional means "if A is selected, B must be selected and if B is selected then A must be selected". So, A → B as well as B → A are possible.

1) Which of the following pairs of animals can go for the walk?
    A)   Anteater, Bison >> correct
    B)   Bison, Cheetah >> incorrect, violates rule B → A
    C)   Cheetah, Anteater >> incorrect, violates rule A → B

4	**GROUPING**

Three animals, an Anteater, a Bison, and a Cheetah..
If the Anteater goes then the Bison cannot go >> A → ~B.

1) Which of the following pairs of animals can go for the walk?
    A)   Anteater, Bison >> incorrect, violates rule A → ~B.
    B)   Bison, Cheetah >> correct
    C)   Cheetah, Anteater >>correct-if A goes, B cannot go, but C can go.

Answers
© Gift Of Logic, Inc * Copying prohibited

5	GROUPING

Three cakes need to be selected from A,B,C,D

1. Write the valid combinations of cakes that can be selected.
    ABC, BCD, CDA, DAB

6	GROUPING

Three cakes need to be selected from a set of four cakes, named A, B, C, and D. Cakes B and C must not be selected together. >> ~BC

Note that this representation means that any combination of B and C cannot be selected regardless of the order in which they appear.

1. Which of the following are valid combinations of cakes that can be selected.

   A) A,B,C >> violates ~BC
   B) C,B,D >> violates ~BC
   C) C,D,A >> valid
   D) D,A,B >> valid

Answers
© Gift Of Logic, Inc * Copying prohibited

7   GROUPING

Four students, Amber, Brian, Calvin, and David.. >> A,B,C,D

If Amber is in the blue team, then Brian must be in the Green team.
>> A-Blue → B-Green

If Calvin is in the blue team, David must be in the Green team.
>> C-Blue → D-Green

Apply these rules to each choice to answer the question.

1) Which of the following team selections are possible based on the above scenario?

Blue team	Green team	Possible?
Amber, Brian	Calvin, David	No, violates A-Blue → B-Green
Amber, David	Calvin, Brian	Yes
Amber, Calvin	Brian, David	Yes
Calvin, David	Amber, Brian	No, violates C-Blue → D-Green

Answers
© Gift Of Logic, Inc * Copying prohibited

8 GROUPING

Four students, Prince, Queenie, Raj, and Sam.. P,Q,R,S

Exactly two students in each room.
Queenie should be in the green room. >> Q-Green
Raj should be in the blue room.>> R-Blue

Apply these rules to each choice to answer the question.

Blue room	Green room	Possible?
Prince, Queenie	Raj, Sam	No, violates Q-Green
Sam, Raj	Queenie, Prince	Yes
Queenie, Raj	Prince, Sam	No, violates Q-Green
Raj, Prince	Sam, Queenie	Yes

Answers
© Gift Of Logic, Inc * Copying prohibited

9	GROUPING

Four students, Amber, Brian, Calvin, and David ..>> A,B,C,D

Amber and Calvin must be in the same room. >> AC
Brian and David must be in the same room. >> BD

1) Which of the following can represent valid room occupancy based on the above scenario?

Blue room	Green room	Possible?
Amber, Brian	Calvin, David	No, violates rule AC
Amber, Calvin	Brian, David	Yes
Brian, Calvin	Amber, David	No, violates rule BD
Brian, David	Amber, Calvin	Yes

10	GROUPING

Four professors, A, B, C, and D are to be seated ..
Professors B and C cannot be at the same table >> ~BC
Note that the above rule applies to both the blue and green tables.
Apply the rule and evaluate the possibilities.

Blue table	Green table	Possible?
A, B	C, D	Yes
B, C	A, D	No, blue table violates ~BC rule.
D, A	C, B	No, green table violates ~BC rule.

Answers
© Gift Of Logic, Inc * Copying prohibited

11	GROUPING

Four professors, A, B, C, and D..

Professor A or Professor B can sit in the blue table, but not both.
>> A⊦⊦B @blue
Note that "exclusive or" is indicated by "but not both".

Blue table	Green table	Possible?
A, B	C, D	No, violates the A⊦⊦B@ blue rule
A, D	B, C	possible, does not violate any rule
B, C	A, D	possible, does not violate any rule

12	GROUPING

Four professors, A, B, C, and D ..
The rules can be represented as follows.
    A⊦⊦B @blue
    B⊦⊦C @green

1) Which of the following can represent valid table occupancy based on the above scenario?

Blue table	Green table	Possible?
D, A	C, B	No, violates B⊦⊦C @green rule
C, B	B, D	No, B cannot sit in two tables
B, A	C, D	No, violates A⊦⊦B @blue rule
B, D	C, A	Yes, does not violate any rule

Answers   152
© Gift Of Logic, Inc * Copying prohibited

13 GROUPING

Four professors, A, B, C, and D..

Professor A must sit in the blue table and professor B must sit in the green table. The rules are represented as follows:
  2 per table, A@blue & B@green (note the &)

Blue table	Green table	Possible?
B, A	C, D	No, B must sit in green table
C, A	B, D	Yes
C, D	A, B	No, A must sit in blue table
D	B, C	No, there must be two per table

14 GROUPING

Four professors, A, B, C, and D..
The rules are represented as follows: 2 per table, A@blue || B@green (note inclusive or, which means both are possible, but just one of them needs to be true).

Blue table	Green table	Possible?
A, B	C, D	Yes, satisfies A@blue || B@green rule.
C, D	A, B	Yes, satisfies the A@blue || B@green.
A, C	B, D	Yes, satisfies the A@blue || B@green.
B, D	A, C	No, A is not in blue, B is not in green. Violates the A@blue || B@green rule.

Answers
© Gift Of Logic, Inc * Copying prohibited

PATTERN PERCEPTION

Question#	Answer
1	A
2	C
3	B
4	A
5	A
6	B
7	B
8	A

FIGURE FORMATION

Question#	Answer
1	B
2	A
3	A
4	B
5	B
6	B
7	A
8	B

PAPER FOLDING AND CUTTING

Question#	Answer
1	A
2	C
3	A

FIGURE MATRIX

Q#	Ans	Reasoning
1	A	ball and bat are used in sports; stethoscope and syringe are used in medicine
2	A	head is covered with hat; leg is covered with socks
3	A	nails are cut with a nail cutter; hair is cut with a scissor
4	A	car transports a few people, bus transports a lot of people; boat transports a few people, ship transports a lot of people
5	B	all the three items are used for cutting; knife is also used for cutting
6	A	crab, crocodile, and duck are amphibians; turtle is also an amphibian
7	B	hat, wig, and crown are worn on the head; a helmet is also worn on the head
8	A	stethoscope, thermometer, and syringe are medical instruments; a sphygmomanometer is also a medical instrument

RULE DETECTION

Question	Answer	Question	Answer	Question	Answer
1	A	3	B	5	B
2	A	4	A	6	A

Answers

NOTES

GIFT OF LOGIC™
Certificate

THIS IS TO CERTIFY THAT

_____
(your name)

has completed the exercises in Workbook-8 of the
Critical thinking & Logical reasoning Series,
thereby gaining proficiency in Verbal,
Analytical and Pictorial reasoning.

_____   _____
teacher/parent   date

NOTES